Workplace Health Promotion

A salutogenic approach

by

Anders Hanson

Bloomington, IN   Milton Keynes, UK

AuthorHouse™
1663 Liberty Drive, Suite 200
Bloomington, IN 47403
www.authorhouse.com
Phone: 1-800-839-8640

AuthorHouse™ UK Ltd.
500 Avebury Boulevard
Central Milton Keynes, MK9 2BE
www.authorhouse.co.uk
Phone: 08001974150

© *2007 Anders Hanson. All rights reserved.*

No part of this book may be reproduced, stored in a retrieval system, or transmitted by any means without the written permission of the author.

First published by AuthorHouse 6/7/2007

ISBN: 978-1-4259-9726-7 (sc)

Library of Congress Control Number: 2001012345

The cover illustration is a part of Raphael's painting "The School in Athens" (1509-11)

The two central figures are Plato on the left and his pupil Aristotle, two of antiquity's greatest philosophers. Plato points upward to the world of ideas. He held that ideas and philosophical knowledge form the starting point for all other knowledge. Aristotle's stretched out hand emphasises that the forms are to be sought in things and that science has both a theoretical and practical side to it where the latter seeks to provide a guide to action.
(Filosofilexikonet, Forum, Stockholm)

For more information about the book: www.salutogenic.se

Printed in the United States of America
Bloomington, Indiana

This book is printed on acid-free paper.

Contents

Foreword   xi

Preface   xiii

Introduction   xvii

   The plan of the book xx

   Explanation of certain key terms. xxiii

1. Working life's new conditions   1

   The workplace, human beings and health 1

   Putting human beings first. 3

   The triangle of interested parties 7

   Customer-led management and the role of human capital 9

   A balanced organisation? 16

   Health- an issue of strategy 18

   Health promotion as strategy 21

   Summary 23

2 Health work- three strategies   25

   Curing illness 27

   Preventing illness 34

   Promoting health 42

   Summary 46

3 Explaining what health is   49

   Ill-health: the lot of human beings throughout the ages .51

Health as a state and resource 52

Dichotomy or continuum? 56

The two dimensional model of health 59

Summary 60

4. Health promotion- a long tradition   63

The historical background 67

WHO established in 1948 68

LaLonde report 1974 .69

Ottawa Charter 1986 .72

European network for health promotion in working life .81

Summary 94

5. From idea to theory   99

Theory at different levels 100

Summary 110

6. The mystery of health and sense of coherence   113

Who was Aaron Antonovsky? 115

Antonovsky's salutogenic model as a theoretical guide for health promotion 116

The parts and the whole 123

Testing a theory's validity 124

From "global orientation" to everyday level 128

SOC as a temporal dimension 137

SOC as a pedagogic model 140

Summary 142

7 Health promotion and its fields of knowledge   143

Specialist or generalist competence 144

Delimiting the field of knowledge 146

Knowledge perspectives 148

Knowledge of health and health promotion 149

Knowledge of human beings and their living conditions 150

Knowledge of processes to bring about change 158

Academic knowledge 164

Several subject areas - old and new 177

Summary 177

8 Health promotion in practice   179

Why health promotion? 182

Is there some "best practice"? 188

Health Promotion as three questions 191

Four criteria for health promotion 194

What distinguishes salutogenic health promotion from promoting health in general? 198

Summary 200

9 Focus on measures promoting health   203

The holistic viewpoint 208

A case of tunnel vision? 213

Focus on promotion fits in with the development tradition 216

What is to be promoted? 217

Choice of level for initiatives. 220

The consequences of promotion 222

Summary 223

10 The workplace as setting 225

From focus on the individual to focus on the context . 226

A setting with several interested parties. 229

Co-operation. 230

Every workplace is unique. 232

System viewpoint 235

Integration 239

Differing preconditions for health promotion 252

Summary 254

11 Conditions governing participation 257

The concept of participation within health promotion . 257

The concept of participation in working life 259

Why participation?. 261

What is participation? 268

How is participation created? 277

Aspects of participation 287

Too much participation?. 291

Summary 293

12 Process orientation — 295

Why process? 296

The structured human-related process 308

Process management 316

Subprocesses 319

Following up and evaluation of processes 322

Summary 324

Afterword — 327

Bibliography — 331

Footnotes — 339

Foreword

When all is said and done, what is the difference between health promotion and the prevention of ill-health? This may sound like an academic question, but it is of great concern to practitioners. Ill-health is largely an issue for which we in the medical profession have been responsible. Preventing ill-health is therefore something where the medical profession can play an important role since it is we who have the scientific instruments to determine the causes of ill-health. Promoting health on the other hand, is partly a matter of things entirely outside those which physicians are trained to deal with, and for which they are responsible. The present book contains an analysis of the concepts which are used in the field of health promotion and also an account of their emergence in an international perspective. The focus is on how we create a workplace which promotes health.

Anders Hanson has chosen to employ Antonovsky's concept of Sense of Coherence (SOC) to create an overarching framework for his discussion of health promotion in the workplace. SOC can be seen as connecting the organisation and character of the workplace on the one hand with the individual's health on the other. The three subsidiary concepts of meaningfulness, manageability and comprehensibility are discussed within a workplace perspective. If the employees' SOC is weak, it is necessary to find out why and to try to improve it since *inter alia* their health will also thereby become better. Hanson investigates which workplace conditions can promote SOC and health. He points out, for example, the importance of a reasonable balance between demands on the one hand and the possibilities of influence and support on the other, as well as the importance of a good balance between effort and reward.

On all this, there are already a number of other books dealing with the subject. But in addition, Hanson discusses how one should deal practically with these issues in the workplace. Moreover, he does it in a manner which is both thorough and lucid. This makes this book very suitable as a course book for students who require learning about health promotion in the workplace.

The arguments in the text are presented in depth and instead of merely a series of assertions, the author points out both the possibilities and difficulties in health promotion work. For those of us who have been preoccupied at a practical level with questions of the working environment, it is clear that we need to arouse the interest of both employers and employees in these issues. Anders Hanson's book is well adapted to this purpose.

Professor Töres Theorell
National Institute for Psychosocial Medicine

Preface

The original Swedish edition of this book appeared in 2004. In order to meet current international interest in the salutogenic perspective in Workplace Health Promotion, the original text of the book has now been directly translated into English.

Workplace Health Promotion is an area which is under rapid development in the industrialised world. The smooth functioning of working life and its attainment of good performance results hinge crucially on people's health.

How are we to succeed in creating a real improvement in the quality of working life? A great deal of effort has been put into finding appropriate and effective strategies for the practical solution of this problem. Among other things, the book sets out to show how the salutogenic approach and employee participation can help us to succeed in the work of health promotion in working life.

In the Scandinavian tradition, two major sources of inspiration for Workplace Health Promotion have been the World Health Organisation (WHO) and the European Network for Workplace Health Promotion (ENWHP). In recent years, there has been intensive development both in the academic sphere and in working life. Since the appearance of the original version of this book, it has become an important source of knowledge and inspiration in the further development of a more explicit and dedicated salutogenic approach to Health Promotion in Sweden. Today, it occupies a place of central importance in the scientific literature for educational courses in Public Health and Health promotion.

The book was been written after many years of reflection about questions relating to health promotion. As early as 1976, when we began one of the first Fitness and Wellness courses at Hjälmared Folk High School near Alingsås, those of us in the teaching group had lively discussions about the factors explaining health and its significance for both individuals and organisations. When I finally got hold of Aaron Antonovsky's book *Unraveling the Mystery of Health* in 1993, it helped me to understand the health promotion perspective and proved an important source of inspiration.

From 1994 my time spent with the teaching team at Vänersborg University College and our many long and constructive discussions also proved highly valuable. At Vänersborg, the presence of both dedicated health studies teachers and critical theoreticians gave our discussions great intensity.

During my time spent on assignments at other workplaces and in teaching students in the health promotion program at Trollhätten-Uddevalla University College, (now University West), I felt an increasing need to write this book. Senior managers and others responsible for personnel matters want to know more about the nature of health promotion. Students demand more knowledge about how one tackles health issues in organisations and working life. All these contacts have provided inspiration.

Special thanks must be given to Joseph Schaller of the Department of Psychology at Gothenburg University where I studied for a number of years. When I used SOC as a theoretical model to study health issues in the workplace, he encouraged me to continue developing my ideas after I had completed my essay on this approach.

There has also been a tremendeous support in my writing from colleagues in the Swedish Health Promotion Group, Jennie Andréen, Staffan Wåhleman and Jonas Brandström. Moreover Jennie has read every line and has provided highly constructive criticism.

I am also very grateful to Professors Owe Petersson, former Head of Health Promotion at the European Office of the World Health Organisation and Töres Theorell at the National Institute for Psychosocial Medicine in Stockholm, who have critically examined the manuscript and at same time have given me a great deal of encouragement in my work.

A special thanks also to Craig McKay in Uppsala who has spent five months during 2006 translating the book. It has been a period with many discussions to find relevant formulations and get the best understanding of the concept in his language.

Apart from the book's starting point in WHO texts, there has been no attempt to adapt its contents to the various traditions and conditions which apply in other countries and circumstances. It is up to you, as reader, to absorb the ideas and experiences presented here and then to transfer and apply them, duly adapted to your own particular circumstances. Many of the scholarly references to various Swedish publications which appeared in the original, have been preserved unchanged in the English edition.

Last but not least, my family, consisting of a wife and four children who have had to put up with a Dad who is too often engaged in sitting, reading and writing, deserves once again a special mention and thanks.

Alingsås, Sweden in January 2007

Anders Hanson

Introduction

The present book sets out to describe health promotion as an idea, a field of knowledge and a strategy for health measures, primarily in working life. One reason for writing the book is the need to have a more unified view of the ideas, disciplines and experiences to be found in this field This overall view provides a useful theoretical survey in helping us to understand what we, as practitioners, are doing. Simultaneously it serves to introduce and develop a relatively new approach to thinking and working with health issues. An exposition like the present one is also needed in order to spread knowledge about health promotion among those who are interested in this approach.

New knowledge does not consist merely of totally new scientific discoveries. Often new knowledge in a particular field arises through a slow process where earlier knowledge is applied in new contexts in a process involving feedback between theory and practice and between researchers and practitioners. The present book can be regarded in this light. It offers, not so much new research, as a proposal for how we can describe and understand what exactly is meant by health promotion.

For many years, there has been great interest in- as well as a great need of-finding alternative ways of tackling the issues of illness and health. Since the end of the 1990s, not least in Sweden, the dramatic rise in absence from work due to illness has increased the pressure to find solutions.

It is not only in the political arena and in the social debate about health issues that there has been growing interest in health promotion and preventive methods for avoiding illness. Colleges, universities and other institutions in the educational sector also want to take part in extending our knowledge and applying new health strategies. There has been a large expansion of courses and training programs within areas such as health education, public health etc. The acute need for a course book like the present one has been brought home to me when I have had to deal with these subjects in various connections.

Since there is a great interest in the idea of promoting health, there is also a considerable literature devoted to this task. A great variety of books are published, ranging from "how to keep healthy" books to doctoral dissertations. My impression is that although the material on offer is voluminous and wide-ranging, there are very few publications in the field of health promotion which provide an in-depth treatment of the subject. Quite simply, what is written tends to reflect about health from the perspective of the individual and deals with what we should do to feel well and about how we should arrange things in a workplace to preserve good health. There is less written about the mechanisms which lead to a change or a transition from the prevailing state of affairs to one which is more favourable from a health standpoint.

The present book provides a more substantial theoretical basis and embodies a three-stage model:

1. The idea of promoting health

2. A theory about *what* is good for health

3. A theory about *how* the change to more favourable circumstances for promoting health can be accomplished.

My starting point is the concept of *salutogenesis*, which forms the conceptual and value-based core of modern health promotion. Instead of compiling a long list of factors which promote the health of individuals or organisations, we make use of Antonovsky's theory of Sense of Coherence (SOC) in order to explain, at a more general level, how we can discover properties, patterns of behaviour and circumstances which are important for health: in other words, an answer to question (2) above. This does not mean that good health advice is superfluous. But since people and workplaces differ, the metatheoretic general model can sometimes be more useful than very specific advice.

Apart from examining the problem of *what* is good for health, perhaps the most important task of the book is to propose an answer to the *how* question (3) above. Among the great mass of literature on health and health promotion work, there is a lack of more general explanations of how we can bring about a state of affairs where there is a lasting improvement in health. It is here that the concept of health promotion has a role to play. Within the international tradition there are ideas and experience originating, above all from the World Health Organisation (WHO), dating from the end of the 1940s, and from EU co-operation since the 1990s, which can help to provide an idea or strategy for health promotion in, for example, the workplace. Much of this has been translated into Swedish and also applied in Swedish workplaces, but consists of efforts scattered over a rather long period of time, which has meant that the results obtained have had only a limited circulation outside political, scientific and other institutional arenas.

It is my hope that the present book will provide an account of the knowledge that has been accumulated, as well as giving a conceptual model which can serve as a guide for future work in discovering effective strategies for promoting the health of human beings.

The plan of the book

Chapter 1 is an introduction to the field of working life and health. The starting point is the change which has occurred - and is occurring - with regard to organisation and working conditions. An imbalance has arisen between the demands of work and people's needs and capabilities. Since society is built on the assumption that we have a productive working life where people can achieve things and feel well, the promotion of health has become a strategic issue. Boards and management must arrive at a balance between the interests of owners, customers and employees respectively, in such a way as to promote a sustainable working life. In the public sphere, a balance is sought between the public authorities (whether at national, county or municipal level), the users and the employees.

In *Chapter 2*, we make clear that we have in fact at our disposal three complementary strategies in health work: namely to cure illness, to prevent illness and to promote health. These three strategies involve two different ideas, several disciplines and a multitude of professions, each with its own methods. The present work deals with health promotion, but stresses that co-operation between different approaches and different professions is required to attain an optimal result with regard to health in general.

Chapter 3 critically examines the concept of health as a step to understanding the quite different conditions which apply in the case of health promotion, illness prevention and cure/rehabilitation respectively.

Chapter 4 provides a historical and international survey of the origin of what we call health promotion. A great deal has been written and published, in many different connections, about the tradition in health promotion and workplace health promotion. In the present book, we have chosen to examine a number of reference works which belong to the main current of health promotion and which, at the same time, provide a clearer basis for the subject.

Chapter 5 discusses the concept of theory from various standpoints. The book has a dual purpose: both to give a deeper theoretical treatment and at the same time to be of practical use. It is not a textbook in methodology, but it can hopefully serve as an orientation in health promotion work for those who value a logical link between, on the one hand, the idea they believe in and on the other hand, the way in which this idea is applied. Theoretical understanding is therefore necessary as a bridge between idea and practice.

Chapter 6 gives an account of the main features of the work of Aaron Antonovsky and, above all, of what led to the theory of a sense of coherence (SOC), a theory which has been both admired and misunderstood. It is often cited and used as an explanatory sociological model, but sometimes it is also misused as a psychological tool. In the field of health promotion, it is the metatheoretic tool which has shown itself to be helpful in finding the factors favouring well-being in e.g. the workplace.

Chapter 7 provides a broad survey of the various disciplines which can contribute to health promotion work. In this domain, the social and behavioural sciences predominate. Health promotion must largely be based on this type of knowledge, but it also perhaps mirrors the author's own background. On the other hand, I do not claim that the disciplines discussed, are sufficient in themselves, or enjoy some kind of monopoly.

Chapter 8 introduces the book's main idea, namely of describing health promotion as a strategy or an approach in dealing with health issues according to a salutogenic perspective. Salutogenic health promotion is defined by four criteria derived from the international tradition. The remainder of the book deals with these four criteria.

Chapter 9 discusses and critically examines the health promotional viewpoint.

Chapter 10 treats the workplace as a setting. What this means, is simply that the work of health promotion is restricted to a given context or situation and takes account of the specific conditions which apply to it. In the present book, it is the workplace/organisation which forms the setting.

Chapter 11 discusses various aspects of participation. This concept is central to most of the international literature dealing with health promotion. The possibility of participation and the wish to do so, also form a very important criterion both for health promotion and indeed for any work attempting to bring about change. It is after all in the interaction between human beings and the individual's own conceptual and imaginative world and attitudes that one finds both the energy for the efforts required to bring about change and the explanation for why people are healthy.

Chapter 12 describes the conditions for a process-oriented approach. Because sound preconditions for health cannot be delivered quickly by experts, but are based on people's co-operation, it is necessary to be able to understand and influence the human-related or so-called 'soft' processes of an organisation in a suitable way. Health promotion is a long-term process of change which is based on learning and a collective understanding both of the determining factors of health and the organisation's goal and the conditions under which it operates.

I have tried to give a clear and comprehensible exposition and at the same time I have sought to avoid value-loaded assertions about how something should be described or how it should be done. The reasoning in the text oscillates, on occasions, between two opposing poles which are typical in scientific and other "official" contexts. Some examples of dualities or opposites which arise in the health area are given in the table beneath. It is up to the reader to decide if the presentation is objective or not.

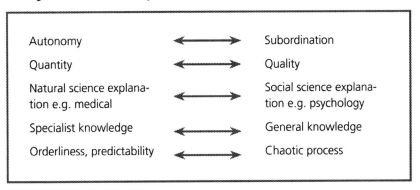

Dualities in the field of health promotion

Explanation of certain key terms

In the present book, the term **ill-health** is used as collective description of a variety of states which cover injury, illness, dysfunction, feelings of discomfort or other states which fall within this category

Public health is the collective - often statistically described - state of health of the whole nation or part of it.

Public health work is the term given to social measures to improve public health. It covers both measures to prevent illness and measures to promote health.

Fitness and wellness program is the term given to activities promoting health which affect people's habits and lifestyle and which presuppose the active participation of people themselves.

There is a debate over the definition and focus of the term **health promotion**.

The concept of health promotion first became firmly established in the 1970s, but the ideas and meaning behind it, date from the 1940s. Roughly speaking, and in accordance with the aims of the present book, we can distinguish three phases in the evolution and extension of the meaning of the term.

Health promotion (Phase 1, 1940-1970). Health promotion as originally conceived,- what might be called *original health promotion* -was essentially a pathogenic approach which looked at the problem of health from an illness perspective. The idea for a broader view of health and health work can be found in the World Health Organisation (WHO) Constitution where emphasis is laid on the need to employ several strategies, namely *control of disease*, *protection of health* and *promotion of health*. Health is dependent upon physical, mental and social factors but the larger environment of our lives is also important. The underlying perspective is the cure and prevention of illness, that is to say, it is pathogenic in character.

Health promotion (Phase 2, from about 1970 onwards). Health promotion in this stage- it may be called *traditional health promotion*- adopted a combined strategy involving both illness-based (pathogenic) and health-based (salutogenic) perspectives. However, there was increasing emphasis on the importance of improving the living conditions and lifestyle of people in order to bring about an improvement in their health. This was underlined in WHO's global strategy "Health for all" in 1981[1], which in fact implied a paradigm shift in moving the focus from disease prevention to health promotion.

Health promotion (Phase 3). In this phase, which deals with now and the future, it is a question of establishing more of *salutogenic health promotion* as a complementary strategy to traditional health promotion with its more pathogenic orientation. Taken together, they form what might be called *comprehensive health promotion*.

The aim of this book is to describe and put forward a more salutogenic health promotion and how we can develop this in working life as workplace health promotion.

1. Working life's new conditions

The workplace, human beings and health

What role does the workplace play in people's lives? What are the requirements which apply to a working member of society in working life? Why are people today ill? Why are they unable to cope or choose to devote their time to something other than their work? Is it the conditions or human beings themselves which have changed so greatly?

Many people have paused to reflect about such issues. It is after all true that we wish that everyone was able to work at something which gave meaning and content to their existence. The work should be reasonably challenging and match the individual's personal competence and other requirements. The workplace which brings several people together to function as a team, should not simply be dedicated to production, but should also allow employees to experience job satisfaction, pride and a sense of well-being. Increasingly the various representatives of working life stress the importance of the link between staff conditions and the results which are achieved. Today it is beyond doubt that a good working environment leads to greater sense of well-being which in turn affects attendance at work, performance and quality.[2] As a result, the costs for absence due to illness and rehabilitation are also less. We know this from the many investigations[3] which, over a period of several years, have registered the increase in absence due to illness and its cost.

We also know that the figures for absence due to illness can differ greatly from workplace to workplace. Differences occur depending

Chapter 1

upon the sex of the individuals concerned, the size of the workplace and upon whether the workplace belongs to the private or public sector. By contrast, we know less about why certain workplaces have very high rates of absence due to illness, whereas others have very low ones. Many studies have explored this[4], discovering different patterns and offering different explanations of the connection between work and a person's health. It may be said that we know - at least at a general level - the significance of a good working environment. We also know pretty well the form that organisation and leadership should have, if it is to contribute both to efficiency and people's well-being.

This knowledge exists in part at a general level in research, in the literature and sometimes in public discussion. Knowledge of what constitutes a good working environment and good working conditions is also to be found in the majority of workplaces. It is probably not a lack of knowledge which causes bad working environments. It is more about how we succeed in applying that knowledge to specific workplaces. What is often lacking is the translation and adaptation of what we believe to be beneficial, to the specific case, Every workplace has its own people, preconditions, problems and possibilities which must be taken into account when we set out to create conditions for a healthy and sustainable workplace.

The workplace is a place where people are brought together to solve some task or other. All workplaces form part of a wider context or setting. It can be an organisation which contains many workplaces. A workplace can also constitute the entire organisation and deliver its goods and services direct to the surrounding society.

Apart from size and formal status i.e. whether it belongs to the private or public sector, there are many other conditions which affect a workplace. Workplaces in the public sector have a task or mission whereas in the private sector they are inspired by a commercial goal. On this basis, an organisation is devised, the enterprise is established, people are recruited and the work gets under way. Certain workplaces have a long history and stability in their work. In other workplaces, we do not know from day to day the conditions underlying their activities: "At the moment we exist but how will it be tomorrow or in a month's time?"

These and many other conditions affect the workplace and the people in it. The more factors there are which we have to take into account, the more complex and variable the pattern becomes. It is easy to verify that there are in fact no two workplaces which are exactly alike when we take account of all the factors which apply in the specific case.

Are there some common features or patterns which are able to explain how things are and what is happening to the conditions of working life? What perspectives or starting points can help us to describe, understand and in time perhaps plan the organisation of the work, so that we can achieve a working life which is better equipped to promote the health and well-being of people than it is today.

Putting human beings first

In their book *Built to Last*, Collins & Porras[5] describe three "P"s which form the basis of a profitable company, namely *products*, *people* and *profits*. This is not simply valid in industry or commercial enterprises. In every situation that an organisation momentarily finds itself, these three dimensions have to be taken into account and balanced against each other. We can also call them three preconditions or domains which are basic in allowing the organisation to function and in ensuring its sustainability and success. Management and leadership are about dealing in turn with

Chapter 1

each of these three dimensions as optimally as possible. First of all, the quality of what is produced must be high; secondly personnel must enjoy and manage carrying out their tasks; and thirdly, the economic aspects must be dealt with in a sensible way.

To what extent, can we allow there to be an imbalance in the priorities assigned to these three dimensions? Is it possible to assign them some internal ranking?

The Ford Motor Company formulated at an early stage its most important guiding principles. Henry Ford preferred to sell a lot of automobiles at a lower price, rather than a few automobiles at a high price. This gave less profit per automobile On the other hand, more people were able to purchase an automobile and derive pleasure from it. It also meant that more people could be employed at his factories and thereby had the chance of a good income. This account suggests that the industrialist Henry Ford was well aware of the value and importance of people – customers and employees – for his company.

Shook[6] describes how later Ford bosses discussed the three dimensions when in 1983 they set forth their statement *Mission, Values, Guiding Principles* (MVGP). When the C.E.O. then summarised the discussion, he said that there was a great deal of talk about the order to be given to the three Ps – People, Products and Profits. It was decided that people should unquestioningly come first (products second and profits third).

By placing people first, the management considered themselves as creating conditions for ensuring that product quality and the corporation's economic return were optimal. As a result of this, the Ford management did two things: first of all, they decided to set the goal of creating as good conditions as possible for people inside the factories. Secondly, they adopted a standpoint embodying values which attracts people in general. An organisation which wishes to achieve what we call good things, is attractive both to employees and customers: we prefer to work in a place which represents something we think is positive and we presumably also prefer as customers to purchase this kind of product.

The majority of people would certainly agree in preferring values which place human beings first. At the same time, the organisation has to work as a unified system. All three dimensions – business activity/ products, people/employees and customers/ profitability – must be taken into account and optimised.

How does the balance between these three dimensions appear today? How is it in your workplace or in your company? How has it been in the Western world in the course of the last 30 years? The example above is taken from industry but it is also applicable to working life in the public sector. The word *market* then takes on another meaning. Public health care and education, to take two examples, are not steered by market forces. Instead it is political decisions and a public body (whether at national, county or municipal level) which determine what is to be done. The scope for exercising influence is comparatively less in the type of organisation which has to carry out the task. Hospitals or schools have to take account of budgets and other democratically formulated policy documents. As citizens or clients, teaching and hospital staff are able to participate like everyone else when they elect the people who wield the power of decision over the general conditions governing what is done.

Irrespective of whether the workplace is private or public, large or small, there are ultimately certain conditions which are common to all when the work has to be carried out. For example, we can always use the three dimensions above, but with somewhat varying nuances depending on the sector of working life we are speaking about.

When we discuss the comparative significance of the three dimensions, we can also say that it is a question of how interest is distributed. A more robust way of putting it, is to say that it is a question of power. Where is the greatest power to be found, the possibility of exercising influence and of having our personal needs satisfied? Is it to be found in the market? In the organisation acting through its management? Or is it the workforce carrying out the work, which wields that power?

Of these three dimensions, it ought to be reasonable to say, as one said at Ford, that the most important thing needed to ensure that a workplace functions as it should do, is that the people who work there, feel fine. When they do, they work more intensively and the quality of their work is better. People work better when there is scope for job satisfaction, pride in what they do and a feeling of well-being. High performance levels and profitability are dependent on people who have the proper competence, feel motivated and are able to cope with the work.

Deciding whether the explanation that an enterprise performs badly, is a lack of physical and mental capability to cope, or is due to lack of motivation or to health problems, can be difficult. A workplace is generally a complex phenomenon: after all, the more people, the greater the complexity. The workplace considered as a social and technical system only partially functions as a comprehensible and logical unit. We can make decisions and take measures and expect that it will then be as we had envisaged. But a great deal of life in a workplace is impossible to predict or understand. Sometimes unforeseen things happen; people react in a way that we had not expected or the result can be quite different from what we initially thought.

> All human beings have their own motives and what spurs them on, varies from person to person. Optimally these coincide with the activities and goals of the workplace. When this occurs, workplace performance is also optimal.

The triangle of interested parties

A prerequisite is that all members of the workforce feel that what they do, gives them something. The work must have some value for us: there has to be some form of reward. " What's in it for me?" is a natural question that people ask themselves. Afterwards, they decide to what extent it is important for them to get involved and to make an effort.

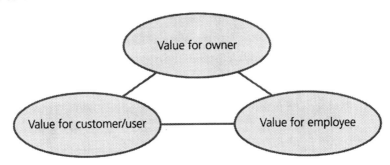

FIG. 1.1 *The triangle of interested parties*

There is therefore reason to reflect about the way in which a workplace distributes the value which is created. What do employees get out of it, how much goes to the organisation and what utility or value goes to the customer or user? These are the three dimensions or the three interested parties weighed against each other in an organised activity which functions well and yields a good result. Even with this description, the demand for a kind of balance between the three remains.

The triangle of interested parties [7] fits in as the description of this interaction both in public and private enterprise. The interested parties which are displayed in the figure above (fig 1.1) are three in number, namely owners, customers and employees.

The *owners* want a return on the capital they have invested. In the public sector, the word *owners* may be replaced by the word *public authorities*. These are the people who have taken the initiative

Chapter 1

to launch the work in question.. The goal is in some sense to create surplus value, a greater return or an increase in social utility.

The *customer/user* is also important. If customers are not satisfied, they do not return. We have to know what the customer wants. It is customer satisfaction and customer experience of quality which decides if they are willing to pay for the goods or services which they purchase.

The employees want their wages. It is also possible to compile a long list of the different values which employees perceive in their work. Maslow's pyramid of needs can provide an example of the various values or needs, work can satisfy.

The triangle of interested parties is a simplification. Nevertheless, it shows rather well how the organisation or workplace forms part of a larger system where different parts of the system are dependent on one another and must work together co-operatively. In reality, there are many more interested parties in relation to a company or an activity. However, it serves as a basic model in order to show the importance of asking more than one group what the work undertaken should contribute to. Is there one of these three interested parties who gains the most from what is done in the workplace? Alternatively we can ask the following question: how do we achieve a good balance when it comes to distributing or apportioning the utility or value which we in our workplace produce?

When we take the triangle of interested parties as an illustration, it is a matter of motives and incentives. The value which is created and apportioned among the interested parties, is useful; it is nice to have and therefore we want it. We can speak of the policy of utility maximisation which each interested party pursues, based on their respective points of view. We can also speak of an optimisation where all the interested parties have their needs jointly satisfied in balance or equilibrium with one another.

There is reason to reflect about the general state of this equilibrium in modern working life. Is there a good balance between the various interested parties? Have we a balanced maximisation of utility? Are the respective interested parties able to have their primary wishes satisfied? The owners want to have a return on their invested capital, the customers want to have high quality services and products and finally the employees, in addition to their salaries, want to experience job satisfaction and to remain in good health.

Since this book begins with the problem of ill-health and deals with health, it is natural to assume that today we have created an imbalance or disequilibrium by concentrating too much on customer satisfaction and the needs of the market. This is well-intentioned and completely logical, since it is at the customer level that value is created in working life which increasingly is devoted to producing knowledge and services. Although it may be goods which are produced, it is in the end customer satisfaction which decides the value of the product.

The fact remains that if there is an imbalance between the various interested parties, then none of the interested parties can attain its optimal value from a long-term perspective.

Customer-led management and the role of human capital

When the management of companies and organisations write their annual reports and policy documents, the latter now frequently contain some elegantly formulated phrase about the great importance of the employees for the results and success achieved: "human capital is the company's most important asset." This reveals the insight that it is in fact the workforce's skills, determination and physical and mental ability to cope, which ensures that the work goes ahead and results are attained. If employees become tired or

Chapter 1

disgruntled, there is a risk that they will put less into their work or move to another employer. Stated quite simply, it is precisely as Leif Edvinsson says[8] : when the working day is done, our most important production capital goes home.

There are lots of signals suggesting that working life increasingly operates under new conditions. Lars Bern in his book on human capitalism[9] present the historical background to the change which has taken place in the capital structure in society. For the main part of our history, from the first human settlements until the industrial revolution, it was land and agrarian capital which was the primary asset. A rich person was a person who owned land. The more fertile the land, the more rapid the process of economic expansion. Since land was a limited resource, growth was also rather limited. As Lars Bern puts it, it takes time to develop methods of cultivation and not all new applications were successful. In the wake of agrarian society, industrial society emerged. Steam and electricity accelerated developments. The most important capital stock consisted in physical capital and factories and machines and the means of transportation became decisive assets.

In contemporary society, a third type of capital, namely immaterial capital, has come to dominate. Bern writes:

> We find ourselves in a stage of economic development in the richer part of the world which has passed the point where immaterial capital has become the dominant form of capital. USA ranks first in this development, closely followed by EU (p.23).

Immaterial capital in the form of bright ideas and a skilled workforce has always existed. The difference from previous times is the great importance that immaterial capital has acquired. Things which originate and exist in people's brains are difficult for organisations and societies to control, store centrally and otherwise

Working life's new conditions

determine. It is a strategic resource which is mobile, capricious and subject to emotions.

When employers in some text or other, show that they are aware that the most important asset is people, it becomes contradictory if staff policy, the working climate and environmental questions relating to the workplace fail to be treated in a way consistent with this line. It is worthwhile to pause and reflect about this idea. Has the equilibrium of the triangle of interested parties (figure 1) been achieved so that the owners, the customers and the employees have received what they need and deserve to have of value and profit in working life?

The picture which has emerged in more recent years, shows that this is most often not the case. There is great attention to the market and profits which has entailed that we have ended up with state of imbalance, where the needs of the owners and customers come before those of the workforce. This development has been clearly manifested in the organisational trends which have characterised working life from the 1980s onwards. There has been a powerful and necessary change towards increased market-led and customer-led management. The global market has stimulated rationalisation and has led to increased demands on the workforce. Priorities, decisions and management are very largely oriented towards the market or the customer and are directed at profits and expansion.

We have created non-hierarchic or 'flat' organisations where each employee is given increased working assignments and has a greater responsibility for seeing that customers are satisfied and value is created. As part of this change, our working lives involve more delegation and more participation, where each employee is given the possibility of influencing their work to a greater degree and is expected to do so. The intention behind this change was a good one. It was meant to be good for both the employee and for the company or employer. The employee acquired increased "freedom" and perhaps more interesting work assignments. The employer got more committed employees and a modern, competitive organisation.

Today we can witness some of the less positive effects which market-led management, customer-led management and a flatter organisation have brought about. People become overstrained and burnt-out, according to certain researchers[10], due to the flat organisation's increased lack of clear focus and the individual's greater "freedom". For many people, work has become less well-defined. Traditional structures and work routines have largely vanished; they have no place in the picture, when we have to be flexible, both as organisations and as people. The boundaries between hours of work and leisure and recuperation have become less clear or have vanished completely. As examples of this phenomenon, we can cite unregulated working hours, flexible time schedules, distance work, extended opening hours and in certain branches round-the-clock opening etc.

The trend towards market-led management, customer-led management and a flatter organisation has taken place at a time when owner interest and economic incentives have grown in strength. Monthly or quarterly reports and the global economy provide examples which are typical of our era. For most workplaces, it is a question of quickly and effectively responding to hard competition. For companies quoted on the stock exchange, stock market pressure to provide frequent financial reports acts both as a carrot and as a stick, as far as their activities are concerned.

In order to keep pace with the new times and the new economic conditions, companies have to adapt and keep up to date. Organisational change and the development of competence have been tools in this process of adaptation. Parallel with this, both the private and public sectors of working life have introduced - or

Working life's new conditions

have tried to introduce - models such as the *learning organisation*, *teamwork*, *management by objectives*, *just in time*, *downsizing* and *continous improvement*.

Taking the triangle of interested parties as our starting point, we can see -above all in commercial life- a pattern for the years from around 1980 onwards. What emerges most clearly is the increased emphasis on the market and customers which has been described above. In second place come the owners who today are completely dependent on the international economy and the global market. As a logical consequence of this, there has been, relatively speaking, a smaller investment in employees. A radical reorientation of the kind we have undergone attracts great attention and resources at a structural level. The response to increasing competition has invariably been rationalisation and greater efficiency in order to allow for a better use of resources. Human beings are a resource and the development has meant harsher work conditions in the majority of workplaces.

Here we shall summarise the key points and underlying ideology of the flat organisation. Those concepts which have become key terms or "buzzwords" in discussions of contemporary economics and working life have been italicised:

(1) The West has the technology to be able to act both *globally* and swiftly.

(2) Industrial society has given rise to great *efficiency* in the production of goods and services. Development and production are carried out at speed. Production is designed to be *just in time* in order to save time and costs. The result is an increasing surplus which is sold in new markets.

(3) International competition arises and those having the best product at the lowest price achieve success in their

sales. This competition is like a tournament where companies make use of their employees as a team and with everyone exerting themselves to be constantly better than their competitors. This is called *continuous improvement*.

(4) In the ever increasing supply of goods and services, customer choice becomes paramount. *Customer-led management* means that the companies require closer and clearer communication with their customers.

(5) Given these conditions, it becomes difficult to work according to traditional hierarchic models of organisation where communication takes time and has to filter down through several organisational levels. Rapid decisions must be made in close contact with the customer. For this reason, we develop a *flat* organisation, so that the number of levels of decision are kept to a minimum In extreme cases and in small organisations, one can make do with just one level: in other words, the person doing the job makes all the necessary decisions. In larger organisations, there are optimally three decision-levels.

(6) In the flat organisation, many functions are concentrated in the work-group or team. We have therefore devoted ourselves to *team development* forms of co-operation and the organisation of the work is slimmed down. Taking a leaf from the world of sport, everyone in the team is expected to be conscious of the overall aim and general goal. The manager becomes a *coach* who gives support to his employees according to their interests and needs. The responsibility for the task in hand is shared by everyone in the team. The goal is important and becomes both something to aim at, an incentive to spur

people on and a measure of success. This idea gave us the *management by objectives*.

(7) Knowledge, initiative and decisions are all to be found in the team. The team develops its capacity through the members learning from one another and from the work itself. This has given rise to the concept of a *learning organisation* as an image of increased knowledge awareness. The development of knowledge has become decentralised through each employee having, in general, more education and training when they begin. The employee's skills are extended and gain in depth in the daily work, in the encounter with customers or users, and in grappling with problems.

(8) To 'flatten' an organisation and allow people to decide more, has a number of consequences. Among other things, there is a redistribution of *power* which is transferred downwards from the top of the organisation to its base. This can cause confusion and difficulties for intermediate levels that may still exist. It can also be difficult for management who have been used to a virtual monopoly in exercising power. Despite increased influence, employees on the workshop floor can also find themselves frustrated if the expectations and the framework, in which they are expected to act, are unclear.

(9) What remains is the formulation of the main goal of the organisation, its *vision*. This has become increasingly important as a motivational factor since orders and programs are replaced by autonomy and situation-based problem-solving.

Chapter 1

Idea-based management, *core values*, and *visions* are differing forms of images by means of which the management can communicate and awaken interest and involvement in the organisation. What appeals to the hearts and minds of employees, creates incentives which other forms of control are incapable of achieving. We are not speaking about clear goals e g about what has to be achieved this week, but about a vision of the future where we neither know *when* nor *if* we shall succeed in achieving it. All we know is that this objective is what we want and we therefore strive to attain it.

These examples can be considered to apply best in business life and commerce. It is also in this sector of working life where new ideas and international influences usually crop up first. The public sector has not lost time in following this lead and has begun to apply many of the ideas that have arisen. Even in the school and care sector, *management by objectives*, a *'lean' organisation* and tea*m-based* responsibility for the work, have been introduced. It has simply taken a little longer to adapt these trends from business life to activities which are centrally concerned with people. In certain cases, the adaptation has not generally been a success, and the public sector authorities have had to rethink their ideas and find other approaches to management. In retrospect, it can be stated that both management by objectives and the 'lean' organisation have been poorly adapted when they were introduced in the public sector.

A balanced organisation?

Today we have a situation where there is a high degree of stress and mounting numbers of people registered ill.[11] How far can this be explained by the fact that working life has become increasingly concerned with market and customers? If this is a major explanation, it implies a redistribution of interest and focus. Ford wanted an arrangement where people come first, then products and finally profit. Our modern working life has landed in a situation where it is reason to ask if priorities have been reversed:

1. Profits (market, economic control, capital assets, shareholders).
2. Product (satisfied customers, quality)
3. People (the workforce)

It is uncertain if there ever was a time when there was a balance between these three parts of the system. However it is probable that we have now rushed into a situation where there is maximal concern for the market and a lack of concern for people in the system. We have introduced new forms of organisation and work which entail increased strain on people, partly physically but above all mentally. The changes bring in their wake a higher pace of change. The interval between each change in the system has become shorter and shorter. Reorganisations or the introduction of new technology has been a recurring feature of the latter part of the twentieth century. This high tempo forms a stress factor in itself.

From a system point of view, we can picture the workplace in a wider context, namely as part of an organisation, a market and a society. By focussing our attention within the organisation, on the workplace and workforce, we arrive at another system level which in turn consists of several parts. At this level, there can also exist an imbalance which explains why the work is perceived as agreeable or soul-destroying.

What about the balance or adjustment between, on the one hand, the technological and structural systems we create to carry out the work in question and, on the other hand, the requirements and needs of the human organism? What human beings, as biological entities, need or can cope with, comes second when new technology or new forms of production are introduced. The spur to development is economic or technological. Innovation gives rise to innovation and the pace of change accelerates. The rapid development has been able to continue, thanks to the fact that human beings are incredibly

Chapter 1

adaptive organisms. They have the ability to draw upon this adaptive capacity when a decision has already been made to introduce new technology or a new organisation. It is easier for me as an individual to adjust, than it is to introduce a modification to the structure of the production system or task. The adaptation of ergonomics, work hours and support from others has taken place during the process and has facilitated human adaptation and sustainability.

Health- an issue of strategy

The introductory account in this chapter aims at giving a picture which at an overarching level can help to explain and understand how we have landed up with the conditions which characterise working life today. How has it been possible for this alarming situation of ill-health in working life to arise in the first place?

Apart from the introductory description, there are many other perspectives which can throw light on the problem and help us to understand. However, we shall content ourselves with the aforementioned perspective as a broad answer to the question of why it has become as it is. This picture can also help to justify the need for attacking the problem of ill-health in working life in a number of ways, and in new ways. Ill-health is linked both to human beings and life conditions and because of this we need to tackle the domain, both in human and structural terms. Health care in working life has long been an issue which has engaged the attention of many different professions. Company doctors, physiotherapists, occupational psychologists, engineers and all the others in the world of Occupational Health and Safety (OHS), have had different roles and are to be found, side by side, in the service of working life and organisations. Today, it is no longer enough for them to exist side by side. The reorientation that has taken place *inter alia* in OHS towards a more interdisciplinary and team-based approach, has to advance and go much further. The actors who play a part in this field require a common language, a shared view about which ideas and approaches are effective, and a clearer commitment to working together with the new workplace health professions and others engaged in this work. It is also time to embody, in a much clearer fashion than hitherto, a workplace health promotion thinking in the everyday work of organisations and companies. It is a question of matching suitable health strategies with the conditions to be found in the particular individual workplace. If the traditional actors in health care and OHS are unwilling or unable to contribute to this, the organisations themselves, for strategic reasons, will be increasingly compelled to take the initiative and seek new ways.

This book is about health: what health is, how and why it is created and preserved, and what conditions are better suited to promoting better health in working life. Health is a quality or asset which is needed if people and societies are to function well. One reason why this book has been produced and why we today devote a great attention to workinglife health topics, is naturally the occurrence of ill-health. The problem is a wide-ranging one and its effects at different levels underline the need for new ideas and ways of tacking things.

In thinking about the strategic importance of the health issue, we can state that the problem of ill-health is to be found at least three levels of society.- the national, the organisational and the human level. At a national level, ill-health is a social economic problem with a high ratio of persons registered ill which places a burden on public funds. There are costs in the form of payments, since more and more revenue is required for sickness payments, rehabilitation and early retirement. In addition there are fewer people in the production process and thus fewer people contributing to economic growth in society. In the worst case scenario, a person is only productive for a few years after a long period of education. The original idea was that every citizen would have a career lasting around 40 years. Discussion and reports in the media present a gloomy picture which leads to difficulties in manning the social sectors which are especially hard hit by ill-health. Young people receive an unattractive picture of a working career.

It is of major interest for investigators and researchers to find out the causes and patterns of absence due to illness. Much is known about how reporting sick is distributed among different groups in society. We also know a great deal about the connection between absence and conditions at work. Perhaps, however, no one is in possession of the whole explanation of the phenomenon of absence due to illness. It is difficult to assemble an overall picture where we can see how the changes in working environment, legislation, the age structure, the state of the economy, the system of public resources, influence the statistics of reporting sick. Is a high level of absence due to illness a price which must be paid in order for us to be able to retain reasonable welfare, full employment, a high rate of production and economic growth?

Another level where the problem makes itself felt, is within companies and workplaces which incur costs when people are off sick. It also affects work negatively when someone is missing. It becomes more difficult for those who remain, to take care of the task in hand. If replacements are hired, it is both a cost and perhaps also a "disturbing factor" during the training period. Recruitment makes demands on time and resources. When key personnel are off work, the effects can be particularly felt. It may prove difficult- even impossible- to replace a burnt-out boss or an employee with special skills in the short term. In the final analysis, workplace health is about how workplaces in the future will be able to cope with their staffing problem. Is there going to be a sufficient number of people with the energy, capacity and determination to cope with the common task in hand? Already today, large organisations[12] are thinking about their supply of skilled staff. It is not simply a matter just of being able to get hold of people in competition with other employers and branches. For companies who are actors in an international market, it is also a question of being able to attract and look after the best employees available. They look for a capacity for innovation, a desire to develop and a sustained capacity for work in their employees. And people like these are therefore a crucial strategic resource for such companies. In the public sector, this is obviously equally true, but with other conditions and consequences.

The "level of the problem" which is most intimately bound up with each and every person is, of course, the situation of the individual. It is at the individual level that things hurt or people say "stop". It is physical, mental, social and economic suffering which is contrary to everything that is meant by good quality of life. As long as we are healthy, we think seldom about it. But when our health begins to fail, we are reminded of its value. Being healthy and having the possibility to work and fulfil our visions about what is important in life, is something incredibly precious.

Health promotion as strategy

Researchers have many explanations about why ill-health arises. We know what risks and issues involving the working environment must be resolved for preventive purposes. Tradition and accumulated knowledge has less to say about the form which work and life should take, in order for people to preserve and indeed improve their health.

The underlying idea in the rest of the book is to describe what it is that characterises health promotion as a strategy geared to improving health. What kind of perspectives, knowledge and methods are there which can help to underline and strengthen the *determining factors* of health.

What we today call workplace health promotion has its origins in a tradition of working to improve health in working life which arose in the 1970s and had its roots even further back in the establishment of the World Health Organisation (WHO) in the 1940s.

Chapter 1

There are different ways of tackling the work of health promotion. In working life, it can be a matter of discovering forms for organisation, leadership and personnel work in general. The long tradition associated with the environmental aspects of work has a self-evident place in this context. Workplace health promotion must come into play at all the levels of an organisation which can, in one way or another, influence the employee's health. The individual retains a responsibility both for his or her own health and that of the workplace. In order to strengthen and preserve health, it is also necessary for the working environment and day-to-day conditions to support the efforts of the individual. Thus health promotion deals with seeking to improve and promote health in a certain setting. In the present book, the setting we focus on is the work place. It is a setting where health and the preconditions for health can be influenced. Although our starting point is a social and organisational perspective, our principal aim is to create conditions whereby individuals can continue to have the ability, determination and energy to cope with their work in a positive way. Human resources need to be constantly regenerated and developed so that they do not fizzle out

What approach should management, employees and health workers adopt in order to create a workplace which provide conditions for all those who work there to experience job satisfaction, a sense of well being and health? How is this plan of campaign to be combined with the goal of maintaining high quality and efficiency in the production process? Seedhouse[13] holds that the international tradition in health promotion lacks theoretical underpinning and its development has consequently been inhibited. But in fact conceptual models and theory are to be found: the problem is perhaps rather that there is too much to choose among. The idea of health promotion is also relatively new and as a result, there has been insufficient research and discussion for the subject to be sufficiently theoretically developed and mature. This book presents health promotion on the basis of an idea, a proposal for a theory to keep us on the right track in deciding what are the conditions which determine health and, above all, how we can manage change in the workplace to bring about better conditions for health.

Summary

The descriptions above are intended to show some conditions which operate in working life and how the situation of people working there is accordingly influenced. We began by speaking of working life in general where owners, customers and employees form a triangle of interested parties. How do we balance consideration for others and the maximisation of utility between these three?

The next point concerned the development by organisations or companies of new technical and other systems. When an enterprise is organised, technology, economics and other structural factors have great importance for ensuring that things are able to get under way at all. Here we can assume that technology and economics have much greater influence on development than the people who staff the organisations.

The third point took up the role of individual employees in the system where the trend has been towards increased democratisation and increased individual freedom at work. Today we can see the effects of the fact that this change has largely taken place without reflection and for this reason, the effects are not always positive. The so-called 'flat' organisation is said to be beneficial both for customers and employees. We have now discovered that it also creates stress for human beings, due to a situation at work which gives them increased autonomy but also makes greater demands.

The task of - and responsibility for- creating a more human working life with healthy and profitable organisations rests on the

Chapter 1

shoulders of political decision makers, owners, management, bosses and employees. How is this to be achieved? In all probability we already possess the necessary knowledge about *what* is necessary. It is rather more difficult to find solutions about *how* it is to be achieved. How are we to bring about an effective process which promotes better health? How do we create a program which involves every employee and is part and parcel of both the workplace and the organisation? The present book aims to help in providing an answer to these questions. Hopefully the reader will find in it ideas that can be used, as well as new knowledge. Above all, we set out to provide the workplace health promotion with a salutogenic approach.

2 Health work- three strategies

Enjoying good health scores highly when people are asked about what they value most in life. Good health, irrespective of the meaning we assign to it, can be seen as a prerequisite for energy and pleasure in life. The human importance of health is also plain from the space assigned to discussing health questions in the press. There one can find tips about how we can get rid of various ailments and what we should do to increase our well-being. Good health becomes a kind of goal or ideal in life. Health is assigned an intrinsic value. A person with good health takes pleasure in being healthy and feeling fit. For many people, health becomes instead a means for achieving other goals. In other words, health has an instrumental value: "it is thanks to my good health that I have the energy to indulge in a host of enjoyable and meaningful activities in my life".

The issue of health has thus great importance for the individual. As a result, it also is significant in all social situations that people participate in. The general health of the population is important for the production of goods and services. There is a need for healthy employees in those organisations which are engaged in producing our material prosperity. But attention is primarily paid to ill-health because it involves costs in terms of absence from work, health care and rehabilitation.

When all these factors are taken together, there are strong incentives for individuals, organisations and societies devoting themselves to activities of various types. In this chapter, we shall sketch three strategies for health which are in operation today.

1. Treat illness
2. Prevent illness
3. Promote health.

Chapter 2

These three strategies arise from two different theories or perspectives. The first two arise from what may be called an illness perspective. The third is based on a health perspective [14] Both perspectives are concerned with defining either ill-health or health.

When we explain health from an illness perspective, health becomes synonymous with the absence of illness. As a result, the illness or ill-health itself becomes an interesting object of study. We must know how the illness expresses itself and what causes it, in order subsequently to develop a cure. It is medical research which is engaged in increasing our knowledge of the link between illnesses and their causes. It is a matter of identifying and describing factors or circumstances which are *pathogenic* (i.e. cause ill-health). Medical science also has access to knowledge about the processes themselves. It concerns itself with the transition from being fit and well to being ill and in analysing why and how this takes place. The name of this process is *pathogenesis*.

In adopting a health perspective, we look upon health as something quite different from the absence of illness. It becomes more difficult to define exactly what health is. As we noted in the previous chapter, health can be described with qualitative expressions such as well-being or as a form of resource. By adopting a health perspective, we concentrate on trying to explain why human beings-despite various types of stress - nevertheless manage to remain healthy. We investigate the factors or circumstances which contribute to increased health, what Antonovsky[15] has called *salutogenesis*.

Every time we employ the concepts or perspectives of health and illness, it is important to understand their relationship. Pathogenesis and salutogenesis are two distinct processes which arise from two very different ways of looking at things. In a theoretical description of these concepts, we can deal with them, one at a time. When on the other hand we address the situation of human beings in real life, we must note that both these processes are present all the time. We can regard it as a tug-of-war between illness-causing and health-promoting factors. At various times, one of them becomes paramount. As a result, the individual finds him- or herself gravitating towards increased health or increased ill-health. If we are to discuss different forms and methods in dealing with health, we cannot omit either of these perspectives. Pathogenesis and salutogenesis are complementary processes in the lives of human beings. As a result, they should also be complementary when we decide how research regarding health/ill-health is to be conducted. They must also be complementary when we decide which approaches should be given priority in practical health work. It is important to cure illnesses and take care of patients. Resources are also needed to be able to prevent the illness, the real cause of which is known to us. At the same time, we have to devote ourselves to finding out what we should do - and how we should do it - in order to preserve and improve health.

The goal ought to be to ensure that each of the three strategies for health can be developed optimally in itself. The types of knowledge and approach required are to some degree shared, but they are largely specific in character. There must therefore be scope for developing our knowledge and method in all three strategies. Every research worker and professional actor ought to regard it as an asset that the three approaches can work together and contribute in different ways to improving public health.

Curing illness

Illness is a threat to a good life. The driving force to survive and feel fit and well has ensured that interest in curing or avoiding illness has always been great.

Chapter 2

The medical science which we apply in the West today has its historical roots in the ancient Greece of 2300 years ago. This cradle of science led to religious and mythological explanations of the cause and cure of illnesses declining in importance and being replaced by a more objective perspective based on observable and quantifiable "facts". The human body began to be studied more systematically and our understanding of various bodily functions increased.

A fundamental assumption underlying medical research and science from ancient times till the present day, has been the *dualistic* view of human beings which was introduced by Plato (427- 347 BC) By regarding human beings as consisting of body *and* soul, it has become easier to deal with their physical bodies as independent systems. This biomedical system could be investigated, described and treated independently without taking into account the effects of psychological and spiritual factors. Plato did not, however, reject the importance of the soul for human health. In one of his dialogues, Plato defines health as a harmony between soul and body. Both parts were important and affected one another. Bodily health was dependent on spiritual health and vice versa.

Hippocrates (460-370 BC), another of the ancient philosophers, is usually called the father of medicine. He also applied the dualistic viewpoint inasmuch as he considered bodily illness as being caused by biological factors. Hippocrates held that sickness was the result of an imbalance in the body and the physician's task was to find some way of restoring the balance. Thanks to Hippocrates, the healing art became medical science where knowledge was based on empirical facts. Under his direction, systematic experiments were carried out in order objectively to describe and understand how the body functioned.

The scientific tradition had a new dawn in Europe during the sixteenth and seventeenth centuries. Before this, views of human beings and nature were very much influenced by Christian theology, in which earthly life was chiefly regarded as a preparation for heaven. This view was replaced by a scientific view of both human beings and nature. These latter were interesting phenomena which were accessible to science and worthy of investigation. The Frenchman René Descartes (Latin name *Renatus Cartesius*) (1596-1650), a philosopher and scientist who formulated influential ideas about the mind, methodology and science, reinforced the dualistic view of human beings. According to him, there is a clearly defined boundary between body and soul and it is only human beings who have a soul. Animals do not possess a soul and their behaviour is completely mechanical. Even Descartes realised that obtaining knowledge about the human soul was problematic. On the other hand, he held that the human body was well suited to being studied with the help of the same methods which were used to study nature in general. Ever since, researchers in the natural sciences have been drawing logical conclusions on the basis of empirical data based on reality. This scientific tradition has been completely dominant for the last 400 years among medical scientists, when they have sought to analyse and describe how human beings function as biological beings. Research has primarily been directed at understanding the connection between illness, the art of healing and health. Faced with each state of illness, the question has been raised of how and with what medical preparation the illness can be removed and health can be restored. This approach which concentrates on removing an illness via biochemical means with the greatest possible precision, is difficult. Aaron Antonovsky, a medical sociologist, who has questioned the role of medicine as the dominant method for achieving health, calls it the "magic bullet approach."[16]

In contemporary society, we invest great resources in steadily increasing medical knowledge and extending our capacity to cure illness. We also often use medical arguments to justify certain

Chapter 2

human behaviour: eat fibre, drink water, give the child tenderness, spend time in nature- they are good for health.

What is it in the individual human being which creates an interest in health and illness? Is it the desire for a good life or is it the fear of illness which determines our behaviour? Sachs and Uddenberg[17] hold that fear of illness is today greater than ever, and that in our contemporary world, the horror of hell of former times, has been replaced by a fear of suffering from an incurable disease. Everyone who works in hospital care recognises the fear which many people have, of cancer and other serious illnesses. Thanks to health tests and check-ups, the health care system can help us to investigate if we have some illness which has not yet begun to give us pain. Sachs and Uddenberg hold that this widespread fear of illness disguises a deep insecurity or uncertainty which has perhaps quite different causes. Medical institutions (hospitals etc) have been built in step with the growth of knowledge and technology and depending on social resources. According to Sachs and Uddenberg, this has led us into a system where every feeling of discomfort or unpleasantness causes us to try and explain this, by saying that we are afflicted with an illness. In the Western tradition, religious, psychological or socio-psychological explanations have had a minor, peripheral role.

Traditional academic medicine's biological and biostatistical approach has meant that we look upon the human being as a mechanical system- a machine. Illness is something foreign which has attacked part of the system. The treatment is geared to repairing the defective part, so that normal function can be restored. Compared to the illness, the patient as a person with feelings and experiences, has a subsidiary role. This has come about because, with the development and accumulation of knowledge, there has been a narrowing of focus among physicians and a specialisation restricted to smaller and smaller parts of the human body.

Alongside this dominant biomedical approach, there are two further medical traditions which have investigated and treated illness. One of these is the approach we call, psychodynamic.[18] It explains illness in terms of psychological and to some extent -opinions differ- in terms of biological models. It is the part of hospital care which goes under the name of psychiatry. It is to psychiatry that we turn when we are afflicted by illnesses of the mind or soul. We are given medicines for relief and cure in the case of for e.g. difficult conditions such as depression and anxiety. In addition, this type of illness or anxiety state can be treated with the help of psychotherapy which lies outside the domain of medical science.

A third research approach which links up with the two previous approaches is psychosomatic medicine.[19] In this approach, one starts from the underlying view that illnesses often have several interconnected causes and that there are links between biological and psychological factors. Both external psychosocial factors and inner psychological ones can give rise to physiological processes which lead to discomfort and somatic illnesses. The psychosomatic viewpoint developed in earnest after the Second World War. One of the earliest and most influential protagonists was the stress researcher, Hans Selye. He studied and described the physiological stress reaction [20] which largely speaking takes the same form in human beings, irrespective of whether the cause is inner feelings or an experience of an external threat or demand. Selye also showed how this stress reaction, if prolonged, can lead to severe somatic disorders. It is due to Selye, that the concept of stress became the link between soul and body which the dualistic tradition chose to ignore.

Medicine and the art of healing is enormously important today. Sachs and Uddenberg hold that we live in a *medicalised* society where great deals of social economic resources are channelled into health care. If this development continues, we run the risk of

Chapter 2

landing in a situation where one half of the population of Sweden will be involved in looking after the other half. This speculation does not perhaps give a reasonable picture of Swedish health care in the twenty-first century. Already in the 1990s, it has become increasingly clear that there are not enough resources to expand health care in the way we would like. A brake has been put on the expansion of health care and instead we have cut-backs.

The question of how we are going to be able to retain and preferably continue to develop a high quality health care system, serving all of the population on equal terms, is not easy to solve. It is not simply a question of organisation. The health of the population- public health- is subject to a large number of factors, some of which are difficult to predict and quantify. There are more and more senior citizens and statistically the older we become, the more health care we require. At the same time, it is unclear how many young people want to train as doctors, nurses etc in the future. The question is thus how the equation is to be solved when the proportion of elderly with health care needs increases while the generations who are to staff the health care system declines. Since the need and demand for care is high, the cut-backs have entailed an increased burden and strain on the personnel still employed in the care sector. Medical care has become one of the working environments most exposed to stress and with a high number of people suffering from illness. What is under discussion is how the staff can cope and how a good standard of care is to be achieved.

One approach which has been tried in more recent years is to transfer the responsibility for various types of medical care establishment to the private sector. Politicians are attracted by a possibility of achieving greater operating efficiency, while the private companies see a commercial opportunity. The private companies maintain that their experience from the commercial production of goods and services can be transferred to the health care sector.

Health work- three strategies

Logistics, financial control and management can be developed in a way that will transform the health sector into a profitable enterprise, without diminishing its quality.

In certain cases, the staff takes over and runs the enterprise. The advantage of this form of privatisation is that the personnel feel a greater sense of participation and increased involvement. As part owners, they are considered to work harder and as a result the enterprise acquires a kind of capital injection which is cost-free in the short term.

In the future, medical care will also still have to deal with a steady stream of patients. As a result, this work will continue to require large resources in the form of money and personnel. At the same time, it will become increasingly urgent to reflect on how parts of the public health system which are not involved in curing illnesses, can be developed. Many Western countries find themselves in a situation where health care must discover new approaches and more efficient solutions. This means among other things that more social sectors must become involved in thinking about public health, so that the pressure on the health system can be held in check.

This insight already existed in the 1970s when the growth of hospital care costs began to take on alarming proportions in many countries. The Canadian Minister of National Health and Welfare, Marc Lalonde, recommended in a report[21] an approach to thinking about the health field where public health would be seen as part of a wider sphere of activity. His point was that that the health of the population was dependent on much more than a well-functioning system of health care. It is, above all, genetic factors, people's life style and the environment they live in which influences their state of health. Lalonde's ideas led to a reorientation in the view taken by the Canadian government about public health priorities. The previous one-sided investment in the treatment of illnesses was

Chapter 2

replaced by an increased investment in preventive measures. It can be said that with the Lalonde report, ideas about preventive measures and health promotion began to figure more prominently in the public discussion. Since then, it has become increasingly common for decision-makers, not only to discuss the care of the sick, but to ask how we can improve preventive measures and health promotion.

Preventing illness

The Swedish Health and Medical Service Act says that " the health and medical service should work to prevent illness". People turning to the health and medical service should in appropriate circumstances be given advice and information about methods of preventing illness or injury" (Act, 1998:1660)(44).

Already in ancient times, attention was devoted to preventing illness and promoting health. According to Plato[22], the art of healing was designed to improve and contribute to human well-being. For this reason, this form of art was regarded as a true art. Besides curing illnesses, the physician's task was to act in a preventive role as well. This was accomplished by means of advice about diet and physical training. Gymnastics, like the art of healing, had a central role in Plato's view of health. The gymnastics teacher, on the basis of his knowledge of how the human body works, could train the bodies of the young by directing them in gymnastics and sports. By means of such training, somatic balance (health) could be maintained and illness could be avoided (prevented).

The tradition of preventing illness would seem to have existed every bit as long as the art of curing illnesses. In Sweden, the work of informing people about health and preventing illness flourished in the eighteenth century. In order to get to grips with infectious illnesses, fines were introduced for offences against local legislation

Health work- three strategies

regarding hygiene. Several of the most famous physicians of the age wrote books about "the preservation of health" "the attainment of the right old age" or "the art of prolonging life" [23]. With the help of preventive measures and progress in medical science, there was success in reducing the number of mortalities at the beginning of the nineteenth century. At that time, one struggled against major illnesses such high infant mortality, smallpox and other infectious diseases. One reason for choosing a preventive strategy, was the awareness that one had to cure many illnesses. By teaching, for example, how to live a sound and natural life, it was considered that illnesses could be avoided and the number of mortalities reduced. Olof von Dalin (1708-63) was one such early health protagonist and teacher. He can be regarded as a pioneer in the Swedish enlightenment with the journal *Then Swänska Argus*[24]. In its pages appeared his poems. One reads as follows:

> "A contented life
> Eat moderately, drink water.
> Entertaining company, sleep at night.
> Work briskly and merrily away, choose your dwelling with care.
> Rest some hour of the day.
> That is the law for my health and my repose".

The work of preventing illness has for many years been a matter of high priority. By taking measures which reduce or render innocuous those factors which cause illness, we allow the individual to avoid suffering and reduce the costs which are borne by society and employers.

An important starting point in preventive work is our understanding of what causes illness. Before we can take preventive measures, we have to know the causes that we have to deal with. Thus just as in the curing of illnesses, it is important to know the

Chapter 2

process of the illness concerned, its *pathogenesis*. It can even be considered more important to know the cause of an illness, since it is this cause which preventive measures sets out to remove. A preventive measure such as vaccination against a certain illness, must be done with great precision, if it is to be effective.

Preventing ill-health can be accomplished in many different ways and with respect to different target groups. A preventive measure can be set in against a specific illness in a particular individual as in the case of the foregoing vaccination example. A preventive measure can also entail the examination of the state of health of a whole population or population group, in order to be able to detect weaknesses, deviations or symptoms which have not yet emerged as illness. An example of this is the check-up carried out on every 4 year old child or measuring the blood pressure of all employees aged 50 or over, as part of company health care. To examine a whole group in this way (screening) provides information about the average state of health in the group and identifies people who run a high risk of illness. Such examinations are of great value to the high risk individuals thus identified who as a result can take measures which can help to prevent their falling ill. For most of the people investigated, however, the test has little or no value since they are completely healthy and run no heightened risk of illness. We should not, however, underestimate the fact that a check-up in itself can relieve people's anxiety.

The socio-economic value can be considerable if the occurrence of "costly" illnesses can be decreased. For this reason, preventive measures are usually carried out at the social level. *Public health work* covers several areas with the primary purpose of preventing illness. It involves among other things providing health information, health policy and health care. The work is carried out at the municipal, county and national levels. Public health is also on the agenda of several EU organs, as well as the United Nations(UN) and the World Health Organisation (WHO).

Health education

Historically, the provision of information about health has attracted the greatest interest as a way of preventing ill-health. Already in 1737, the Swedish parliament set up a so-called "health commission" [*sundhetskommission*] which had the job among other things of spreading information about how people should live in order to preserve their health. The underlying assumption was that individuals themselves were responsible for whether they became ill or not. "Bad habits, sloth, low morals, neglect, abuse, ignorance, obstinacy, bad sexual practices etc were considered to be the underlying causes of the bad health of the population."[25]

During the nineteenth century, Sweden began to be industrialised and the political leadership realised eventually that there was a clear connection between the workforce's state of health and productivity. Good health was crucial in being able to perform well in the often heavy and dirty jobs, involving long working hours, which were characteristic of the industry of that time. As a result, the workforce had to be trained- or according to the views of the age - "brought up" to superior ways of life in order to contribute in building society. There was a campaign for cleanliness, good eating habits, physical exercise and a balance between work and rest.

An important role in Swedish nineteenth century physical education was played by Per Henrik Ling (1776-1839), the father of Swedish gymnastics. In 1823, he established the Central Gymnastic Institute (later to become the Swedish School of Sport and Health Sciences) in Stockholm and devised a system of gymnastics for comprehensively training the body which became the standard model for Swedish school gymnastics, far into the twentieth century.

Chapter 2

A forerunner and source of inspiration for this gymnastic tradition came from the armed forces of the time and traces of it are clearly present today in school gymnastic halls. Information about health and bringing up people to be healthy were thus socially important both for a strong defence force and for a productive labour force in the emerging industrial sector.

One further concrete Swedish example of providing information on health is the work of the district medical officer in Norrland who around 1830 began to campaign for breast-feeding among the Finnish-speaking women in Tornedalen[26]. When these women ceased feeding their children with the help of a suckling horn and began to breastfeed instead, infant mortality sank dramatically in the region. This change brought about by spreading information about health and associated, above all, with the district medical officer, Carl Joshua Wretholm in Haparanda, was no simple process. It was a question of breaking with a long established cultural pattern. Moreover it was not easy for the individual woman to see the connection between epidemiological data concerning infant mortality and changing the way in which she fed her children. By recruiting a trained midwife who enjoyed the trust of the local people, the doctors were able to gain more support for their ideas. Nevertheless it seems to have taken a generation before breastfeeding became the usual way of feeding infants with mother's milk.

During the twentieth century, providing health information has been a self-evident social task. Parliament and the authorities have, time upon time, carried out investigation of population and public health matters. Following these studies, directives have been issued for how the population should be educated to embrace a more healthy life style. As members of society, we have imbibed this health information in various circumstances, for example in school, from the district nurse or through brochures popping in through the letterbox.

Health work- three strategies

Eriksson and Palmblad[27] describe how public health issues achieved prominence in the 1930s as the result of the supposed population crisis. Hard times meant that illness increased and the birth rate declined. With this in mind, it was thought important for national development to give the citizens new confidence. Alva and Gunnar Myrdal published their book *Kris i befolkningsfråga* [Crisis in the Population Issue)] in 1934. The discussion which ensued, led the government to set up a Commission on Population with the task of investigating both population and public health issues.

Olsson[28] has described the conclusions reached by the enquiry. Among other things, the Commission noted that since too few children were being born in the country, it was more important than ever before that *"the new generations grew up to have the greatest possible physical and mental health with consequently increased possibilities of supporting the future productive life of the country.*

However, apart from this, the public health level would be raised through two important strategies. One was the public provision of information and health education by the authorities. It was no longer enough to leave the information and educational work entirely to individual initiative. It was a task for the school, public education etc. School children were regarded as the crucial target group since it was considered easier to influence the habits of younger people than those of their seniors. Children would also have the role of transmitting health information to the home by means of leaflets with advice and instructions.

The second strategy which the Population Commission advocated, involved social and economic reforms. It was considered the task of society to support those who, because of poverty or ignorance, had a poor diet. Among other things, free school meals were proposed for all children, as well as better school kitchen instruction.

Chapter 2

It was important that the informational work was based on scientific facts. With this in mind, the National Institute for Public Health [*Statens institut för folkhälsan*] was set up within the Ministry of Health and Social Affairs in 1938. Its task was to carry out investigations and pursue research in issues of importance for public health, particularly those aspects dealing with environmental hygiene. It was also part of the Institute's tasks to train health care personnel such as doctors and nurses. They, in turn, could pass on knowledge to the general public.

The information which was disseminated, dealt not so much with providing a foundation of knowledge which would allow people to make their own choices, but rather with concrete advice and instructions about what one should do to stay healthy. In school, it was a matter of rules which had to be followed without question. I personally remember how the obligatory scrubbing with stiff brush and soap was supervised by the "bath ladies" at The Central Swimming Bath in Sundbyberg in the 1950s. Hygiene had a very prominent place in child health training.

Public health work today

Public health work has historically been overwhelmingly concerned with preventing illness. The scholarly-scientific tradition, the measuring methods employed, the policy and the focus of activities are all based on an illness perspective. The determination to also incorporate a health perspective in public health work has, however, increased in recent years. Currently in many places, there is also a clear health promotion focus to public health work. It is indeed difficult to find public health schemes today, which do not present a health promotional goal and direction in their statement of objectives. However, the extent to which these plans are put into practice with salutogenic measures varies greatly.

Health work- three strategies

The current trend of public health work geared to health promotion, in Sweden and elsewhere, began in the 1970s. An important milestone in this connection was the Lalonde report[29] which has been previously mentioned (see p. 33). The situation in Canada was similar to that in Sweden. There was an alarming increase of needs within medical care. Lalonde drew attention to the need for increased investment, above all in preventive measures. He held that of the four "health areas" which he described, it was people's lifestyle and behaviour, in particular, which merited the greatest investment.

Although Lalonde advocated an increased investment in preventive methods, his report became the starting point for radical rethinking. Lalonde drew attention to factors such as environment and lifestyle in his report. This marked the beginning of a discussion within the framework of the work of the World Health Organisation which increasingly became preoccupied with what people can do to promote public health - or more exactly to promote those conditions which would foster good health in the population at large. "Health for all" became a slogan which was adopted at a conference which WHO[30] hosted in Alma-Ata in 1978. The declaration was very much concerned with social justice. Afterwards "Health 21" which became the European regional program, was launched along with strategies for implementing the global WHO policy.

Public health work aims at the good health of the population at large. This holds both for preventive measures against illness and health promoting measures. How successful is such work? Are we able to obtain figures about it? Evaluating such efforts turns out to be a little more complicated. If we wish to evaluate the success of the preventive methods at the individual level, it is not really possible at all. We do not know if it was the preventive measures which led to the person not becoming ill. We do not even know if the person concerned ran the risk of becoming ill. It is not possible

Chapter 2

to evaluate or measure an event which has not taken place. If the preventive work is successful, then nothing happens and no one becomes ill.

At the group and population level, it is easier to estimate the result of preventive measures. If all school children are allowed to rinse their teeth in fluoride regularly for a number of years, dental care is able to register that the number of children with caries declines. By comparing groups, the effects of the measure become clear and the statistical results are reliable. It is also at the population level that the measurement of health has great importance. Within the science of epidemiology, society keeps an account of the prevalence of various illnesses and disorders, how they change over time and which groups are specially affected. Research can be concentrated on the connection between various illnesses and a host of environmental factors and living conditions which can be suitably quantified. These illness statistics form an important basis for health policy decisions and the planning of society's investment in health care.

Promoting health

For long periods, health care has consisted of *curative* and *preventive* measures. We now come to the view of health work which is the primary focus of this book, namely health promotion with a salutogenic approach. People may wonder why we should bother to complicate matters by introducing the concept of salutogenic health promotion. After all, it is still the case that if we work in accordance with this approach, there will presumably be fewer ill; what has been accomplished seems superficially simply to be a *preventive* measure, although we choose to call it by the name of health *promotion*. Can we not simply make do with the first two approaches? The answer is 'no': it does not suffice if we hold that it is interesting to understand and apply both *pathogenesis* and *salutogenesis* as explanatory processes. There is need for greater knowledge- that is to say research- in both these areas. *Pathogenesis* and *salutogenesis* are two qualitatively different ways of looking at things which we need to understand and know more about. If we confine ourselves to the curative and preventive approaches to illness, our starting point is *pathogenesis*. We make use of knowledge about the connection between the cause of illness, its symptoms, its diagnosis and treatment. Scientific medicine has built up its knowledge on the basis of this and obviously must continue its research in order to improve curative and preventive methods.

Salutogenesis is quite another idea. In order to study it, researchers must direct their questions in a quite different direction. The task is then to investigate what it is that helps to preserve or improve people's health. In this approach, we focus on the whole human being and the conditions which mould his or her life. As a result, a multidisciplinary approach to the subject is required. When health promotion work is to be implemented, it needs to be linked to the different "settings" where daily human life takes place, e.g. the workplace, the school or local society. When Antonovsky[31] presented his idea of salutogenesis in 1979, he wanted to make clear that there are constantly two different directions we can take and two forces operating on us. On the one hand, there is a *degenerative* process which leads to increasing ill-health; on the other hand there is a *regenerative* process which tends towards improved health. When both these aspects of health are simultaneously present and at work together within us, this determines if our health is preserved, degenerates or improves (i.e. is regenerated.)

Antonovsky helped us to understand the logic of health promotion by means of the continuum model which describes human health as something dynamic with movement and direction. From a given point on the health continuum (the line between the twin poles of health and illness), different conditions or measures can act to improve the individual's health so that he or she is moved nearer towards the health pole. This state of health can be superior, both when measured in physiological terms and when considered subjectively in terms of the feeling of health experienced.

There exists a certain confusion about how the concept of health promotion should be used. As a result, the representatives of the various health professions find themselves sometimes talking about two different things and failing to connect. Some maintain that health promotion includes everything that is done to improve the state of health in the country, and in this case one tends to think primarily in terms of curative and preventive methods. It is certainly the case that one can see things in this way. It is a natural viewpoint since the pathogenic approach with its emphasis on cure and prevention has been the dominant one over a long period. It also fits in better with Western ways of thinking with its emphasis on a problem-solving approach where we try to find logical solutions based on cause and effect. We cure illness as it occurs, or else prevent it when we know its underlying cause. On the other hand, we are always in a position to promote health as a practical method of resistance.

In the international discussion, health promotion has hitherto largely been about curative and preventive methods. This can seem somewhat illogical and strange, but it is because the idea of salutogenesis is relatively new, while the concept of health promotion has a much older tradition, with roots in the 1940s when social reconstruction was very much concerned with bringing about a well-functioning health system.. After Antonovsky had introduced his idea of salutogenesis, interest in it has gradually increased and today both in Sweden and elsewhere, health promotion embodies major components of the salutogenic approach.

The radical revolution in thought brought about by Antonovsky has astounded many people both because of its ingenuity and its simplicity. Despite this, we are only at the beginning of a

Health work- three strategies

development which will eventually provide us with well tested and practical methods for salutogenic health promotion. The care with which we define concepts can seem to have purely academic interest and in daily speech it can play less of a role if we allow the concept of health promotion to cover everything i.e. curing, preventing and promoting. There are, however, three occasions when it is important to be careful with our definitions and to make it quite clear about the meaning of the term we use.

The first is when we are carrying out research and we require to formulate unambiguous questions. The interpretation and textual presentation of research also requires conceptual clarity.

The other occasion is when we are going to develop new approaches together with other professions. If a physician, an engineer, an economist and a health educator are to co-operate in a health project in a workplace setting, they have to agree about what is meant by key concepts. Effective co-operation requires that all the participating parties speak the same language. Otherwise there is a risk that ways of working and the levels at which initiatives are pitched, become confused and muddled in a way which makes the project more difficult. Indeed it may even counteract the goal of the project.

Finally, there is one further occasion which is linked to those two already mentioned, but which is worth a mention in its own right. This concerns evaluation, when a measure or practice has to be assessed. It is important for us to be able to quantify the effects of what we do. Evaluation is a key concept in all work involving development and change. Describing how things were before, during, and after a given process, helps us to see if this approach is effective or not. We want to have a basis for further planning or a account which can be used to convince decision-makers. If we are not absolutely clear about what is being described or measured and about which methods are appropriate, our results will not be especially reliable. Evaluation and so-called applied research are two areas where clarity about the concepts involved and the choice of method determines the reliability of the results we obtain.

There are thus several reasons for discussing the meaning we attach to the concepts we use. By trying to attain a more or less unambiguously fixed definition of what words mean and how they are related to each other, it is then easier to engage in an open dialogue where different professions and viewpoints are involved. It is certainly stimulating and constructive to raise questions about our own pictures and explanations of things. It is quite another matter if it is then possible to reach agreement about interpretation.

Summary

The aim of this book is partly to show that today we ought to conduct health work on the basis of three different strategies which are available. These are:

1. *Treating and curing illness* where knowledge is required about illnesses and their causes. The methods vary depending on the illness involved. The aim is to find a treatment which is as effective and accurate as possible and with the fewest possible side-effects.

2. *Preventing illnesses* is also based on knowledge of what causes ill-health. Certain preventive methods e.g. vaccination are restricted to a particular illness. Other kinds of preventive work can be more indirect in character, with the aim of influencing people's life style or behaviour. Preventive methods can also consist of adapting or limiting risks in people's immediate environment. The effects of preventive methods can only be evaluated in the case of larger groups, or in the case

of a whole population. It is impossible to decide with certainty in the case of an individual if a certain vaccine or a certain type of behaviour has protected that person from an illness.

3. *Health promotion* is here described as a third strategy which aims at finding measures developing health as a resource in everyday life. The salutogenic idea about what it is that allows people to stay healthy, is still rather new. What is there in people's way of life or in them as persons, which functions as a *general resource of resistance* to illness or as factors supporting health? What can be done not only to see that health is preserved but that it can even be improved? With the help of the idea of salutogenesis, this strategy can be developed into an approach of its own which clearly differs from both curative and preventive health work.

3 Explaining what health is

Enquiries have shown that ill-health is expensive for the individual, the workplace and society[32]. As a result, there are strong incentives for carrying out various measures in order to be able to get to grips with the current situation. For a long time, there have been numerous voices advocating an increase in investment in both rehabilitation and preventive work to reduce the number of people who are ill. Other approaches which are being tried, include new rules in the social insurance system in order to forcibly bring about a change in the pattern of registering oneself as sick..

The traditional and predominant methods which are advocated, share the common aim of trying to reduce ill-health. It is therefore also mainly knowledge about the nature of illness and its causes, which forms the basis for decisions about what measures should be taken. Research with its focus on problems constantly expands the pool of such knowledge that is available.

Ill-health is the customary starting point. It is also that which drives us to grapple with the problem and take action against *ill-health*. In recent years, it has become more and more common to employ formulations which stress the importance of investing in improved *health*. This division of the problem into two quite different ways of looking at the problem has several advantages, since it is a matter of two quite different approaches to describing and tackling the issue. The knowledge which is able to tell us what ill-health (illness) is, differs from the knowledge which can tell us what health is.

In presenting objectives, the goal can be described in terms of the minimisation of ill-health. A workplace with a certain absence due to illness can set itself the target of reducing this by half within e.g. a year. As an unrealistic vision, we can aim at eliminating ill health altogether: in other words, zero days of absence due to illness.

If, on the other hand, we start out from what health is, the goal becomes more difficult to formulate. If health is synonymous with people being at their work and not registering as sick, the goal can be to increase the number of employees who are "long-term healthy" [33]. It can be a valuable goal which both reduces costs and in fact implies that the individuals involved are less ill.

Health work can also have a higher goal which goes beyond having as many staff as possible present at their job. If health is described as an everyday resource for all human beings, it then becomes possible to direct our efforts at strengthening this resource. Health work can then aim at contributing to contentment, job satisfaction and to people's well-being. The calculation of the number of days lost to illness or the number of days present at work can then be supplemented by qualitative measures which take account of the individual's personal experience of health and well-being. It is such a goal which must characterise *healthy working life* or *the healthy work place* and which benefits society, the employer and the individual.

When the concepts of *health* and *ill-health* are used by politicians, researchers and health professionals, there is sometimes conflict and misunderstanding. This is unnecessary: both concepts exist and both are needed. It is valuable to have two complementary views which jointly contribute to the growth of knowledge in the field of health, thereby enabling us to work in several ways with curative and preventive measures against ill-health and with the preservation and improvement of health.

Ill-health: the lot of human beings throughout the ages

Throughout the ages, human beings have been exposed to different illness-generating factors. Historically there have been different views and explanations about why people are afflicted by ill-health. In ancient times, and even among certain indigenous populations today, it was held that illness was a punishment imposed on human beings by the Gods for disobeying the divine will. It is not all that long ago since we in Sweden also believed in evil spirits, witches or trolls as an explanation for why someone became ill. If not a punishment, ill-health could be looked upon as a trial or test, designed to inculcate in people the value of longsuffering patience and a preparedness to accept one's lot.

If we go very far back in time, the Chinese 4000 years ago had reflected about the causes of illness. A common explanation in the Chinese tradition is that ill-health and illness depend upon an imbalance between two elements Yin and Yang. The Chinese held that a balance must be maintained between them by living a sound life.

In contemporary Western society, we believe in other causes of ill-health and illness. Through research, we have discovered the causes of illness in a large number of areas. One can become ill through bacteria and viruses or through toxic elements in food and the environment. Illness can arise through inherited genetic factors or through negative effects of our lifestyle. One area of particular relevance, which has long been of interest to many researchers, is the link between negative stress and illness.

Today in the Western world, there is much talk about the illnesses attributable to affluence and what might be called "the alienation syndrome". Illnesses due to affluence are caused by our lifestyle and our unnatural way of using our bodies. The Western world has a surplus of food and most of all, consumes too much fat food. In the case of many people, the nutritional value of their food in terms of vitamins and minerals is far too low compared to the amount of calories consumed. In addition, the food is subject to massive processing and is poor in fibre. This leads *inter alia* to cardiovascular illnesses, obesity and diabetes.

Another crucial factor of lifestyle is lack of movement and insufficient physical demands on the skeleton and muscles. We live increasingly in comfort and do not require to run and lift. It is notably the generation that was young at the end of the twentieth century which has "suffered" from considerably less physical demands than earlier generations. For most young people, television, computers and the automobile have replaced games in the open air, cycling and walking.

Edgar Borgenhammar, formerly professor at the Nordic School of Public Health has minted the concept of *alienation syndrome*[34]. He holds that in contemporary society, there are elements of psychosocial stress in the form of loneliness, meaninglessness, rootlessness, conflict and so on, which lead to human beings suffering, not only emotionally but also in due course physiologically. Research also increasingly shows that the form of stress which is caused by social vulnerability is more serious than many other factors in our life style.

Health as a state and resource

There is a great and ever-growing amount of knowledge about the causes of illness and how illnesses can be cured. Nowadays more and more people are interested in discovering ways of promoting health. As a result we need more knowledge about how and why health can be improved and how it can be preserved. It is not enough to state that health is the same as not being ill. We must also be able

Explaining what health is

to describe health in *positive* terms as a human quality or resource in a person's everyday life. With this in mind, we shall review some concepts and models which allow us to describe what health is, and the relationship between the concepts of ill-health and health.

Not long after the Second World War, the World Health Organisation became preoccupied with health problems and already in 1948, put forward a definition of what is meant by health[35]. According to WHO,

> Health is a state of complete physical, mental and social well-being and not merely the absence of disease or infirmity.

In this definition, prominence is given to a holistic view of man. It can be said that it constitutes the first step in extending the definition of health from a purely physical or physiological one to cover the many aspects of human life. This definition of health encountered criticism because it was based on a value-based and utopian formulation in terms of complete physical, mental and social well-being. At what point can we consider that such a state has been achieved? From a global perspective, it is something which for many people is impossible. It is considered unrealistic to have such a formulation as a goal and even as a vision. Even in a welfare society, the majority of people have aches and pains, feelings and states which more or less affect their well-being, but nevertheless they would describe themselves as enjoying good health. In connection with the WHO conference in Ottawa in 1986, the original definition was reformulated. Human health was described as something more than a certain state. Instead health was presented as a resource in daily life. Enjoying good health is not merely an end in itself. Health also possesses a so-called instrumental value: it is something which can help the individual to achieve other goals.

Chapter 3

There is also another way of describing health, namely that which is based on how individuals themselves experience their situation and their state of health. When we then use the expression *health as it is experienced* or *well-being*, the state of health is based entirely upon the individual's own personal judgement.

Altogether then we have three different ways of explaining what health is.

Clinical Status

Here we find a health criterion on the basis of human physiology and anatomy. What are my relevant values? Are my blood-pressure and other physiological values biostatistically normal? Normality is thus defined in terms of the statistical average for people of a certain age or sex. If a person undergoing a medical examination is found to have good or normal values, his/her health is good according to this definition of health.

Functional ability/capacity

To enjoy good health in social terms implies that health is described as a functional capacity and a resource which allows one to participate in the life of the community, cope with everyday needs and do what one wants. It is a matter of debate whether this functional capacity is to be judged by the individual concerned or by some "expert". Nordenfeldt[36] holds that health consists in *being able to attain crucial goals in life*, which implies that it is the individual's aspirations and goals which are crucial when they decide their state of health. People with high aspirations who are physically limited, tend to rate their state of health as less well.

Well-being

Well-being is a person's overall experience of feeling well or unwell. This experience is, on every occasion, unique for each individual. Although it is the *experience* which is the object of enquiry, it will be affected by how the body "feels" or functions. Both clinical status and functional capacity can play a part, when we estimate our well-being. The German philosopher Hans-Georg Gadamer[37] places less emphasis on well-being as an experience and more on how well-being is expressed in a person's behaviour.

Receptivity to new things, a readiness to start new projects and forget oneself, is indicative of well-being. It can be experienced as logical and natural that illness or lack of well-being leads to some form of introversion and self-preoccupation which is more of a state than an experience. When this is the case, it also seems logical to describe well-being in Gadamer's words as

> "a state which is characterised by being involved, of being in the world, of being with our fellow human beings and friends, of being actively and fruitfully involved in our daily tasks" (p. 266).

The above are three examples of definitions which can be used in discussions about the nature of health. They are different forms of state which can be treated singly or jointly. The optimal state combines all three aspects: a good clinical status, a good functional capacity and a good sense of well-being. In real life, it is obvious that there are many people who have a sense of well-being and consider that they enjoy good health, despite a handicap or a chronic illness.

Chapter 3

Dichotomy or continuum?

In considering health as a state and a resource, we are prompted to other thoughts. From an illness perspective, health is the same as not being ill. There is a sharp line between being ill and not being ill. It is the doctor's diagnosis and clinical tests which decide the matter. In order to receive treatment, the patient must be classified as ill, and therefore has to have crossed the dividing line which is appropriate in the case of the particular illness in question. This is an example of a *dichotomous* point of view where one judges health as an "either-or" phenomenon. According to this outlook, there are only two possibilities: either one is ill or one is not ill according to the criteria which apply. A physician makes her/his diagnosis on the basis of a deviation from the normal, and knows in general at what point treatment is required. This principle is valid for medical treatment where there are tested pharmaceuticals where their effects on the illness are known.

Alternatively we can describe a state of health in terms of a *continuum*, which means that health is seen as a line or axis on which people find themselves placed at every moment of their lives. We move between two opposite poles - at one end complete health and at the other end, complete ill-health. Aaron Antonovsky[38] called it *the health ease/disease continuum*. It can be depicted as in figure 3.1.

ill-health  health

FIG. 3.1 *The health ease / dis-ease continuum*

Antonovsky chose to describe health in this way in his reflections about how we can escape from the traditional dichotomous viewpoint that certain people are healthy while others are ill. By exclusively relying on this view, we find ourselves limited in our approach to health issues. A dichotomous viewpoint makes it natural to choose a treatment aimed at curing the illness of those defined

Explaining what health is

as ill. Healthy people on the other hand can be suitably offered a fitness and wellness program or similar activities which help them to keep fit and healthy, and prevents them from becoming ill.

These two approaches to health work are both, in their different ways, important. The former means curing and alleviating suffering which is a self-evident task for a humane and developed society. Fitness and wellness programs which originally were based on the idea of illness prevention, maintain that their work is an important alternative or complement, since it leads to an enhanced quality of life or reduced health care costs.

Antonovsky held that neither individuals nor groups can be divided sharply into the two exclusive states of being either ill or healthy. This dichotomous view leads to a limited and rigid picture of reality, where people become patients even although it is only perhaps a little part of them which does not function normally.

The continuum viewpoint provides us with a perspective which is much closer to reality. At each moment in life, we find ourselves placed somewhere on the axis between complete health at the one end and illness at the other, which Antonovsky sums up in the words. "Sometime we shall die". An important aspect, that he addresses, is the fact that even those who are seriously ill, are still in some sense healthy. In other words, as long as there is a flicker of life in us, there is a trace or fragment of health. The conclusion which then follows from this is that even people who have been struck down by illness or some form of ill-health, can need and benefit from a response to promote what is healthy. There is always something 'healthy' in every human being to preserve, care for and develop.

When we choose to describe health as a continuum, it has a number of interesting consequences. Antonovsky introduced this viewpoint as a way of working with health on the basis of two complementary approaches. Research thereby can be conducted with

Chapter 3

two sets of problems in mind. Among the questions confronting the research worker is the following: "What causes illness?". The work of clarifying the cause of illness is to focus on *pathogenesis*. The opposite concept, so-called *salutogenesis*, means that instead we investigate what it is that helps to ensure the maintenance of health and even its improvement. What factors function as supports and buffers? What is there in human life which, despite all the things people are exposed to, nonetheless acts to preserve health or even improve and restore health?

A continuum view of health, together with the salutogenic way of formulating the question, is a model for how health promotional work can be defined and designed. To put it simply, one can say that it is a matter of starting from the position on the health continuum where people currently find themselves. The task is then to find responses which can help them to move nearer the health pole.

It can be asked what significance it has for the patient or client and for the health personnel/therapists involved, if we choose a dichotomous or a continuum standpoint. Does the medical treatment and care affect the patient differently, depending on the view of health adopted by the personnel? What significance has it, if the doctor has time and interest to base his judgement on the basis of the patient as a whole human being? Does it have significance for patients if they are placed in a ward for the care of the seriously ill or a ward for rehabilitation? To these questions, there are no clear-cut answers. However experience and research[39] support the following thesis: "If people define situations as real, their consequences will also be real." People who are treated as ill would thus take longer to recover than those who are treated as being in a process of recovery.

Explaining what health is

The two dimensional model of health

The idea of how we can describe these different dimensions of human health in one and the same diagram has been around for some time. The starting point is the use of a so called *four field diagram* by looking upon health as being defined in terms of two variables, represented pictorially by the familiar co-ordinate system. Several variants of the two dimensional model of health have appeared in the literature where *clinical* health (well-ill) is measured along one axis and *experienced* health (well-being) is measured along the other. One of the first to present such a model was Katie Eriksson[40], a Finnish Professor of Medical Care. Rydqvist & Winroth[41] have presented in their book a variant of the health cross with expressions which provide a good verbal understanding of what is involved

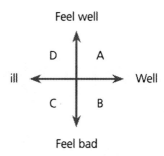

FIG. 3.2 *The two dimensional model of health*

By combining the two dimensions, we then get a figure with four different fields or states which illustrate the importance of taking account of both aspects, when we try to give a comprehensive description of a person's health.

    A. Ideal or optimal state of health where I have both factually good health and experience my health as good. The further up I come in the right-hand corner, the greater my well-being and the less my illness.

> B. In this field, I have at most a few aches but I do not feel well. If this state is long-term, it can lead to physical illness.
>
> C. To be ill and feel bad is the worst state of all. A person with a handicap perhaps begins in this field. When they have gradually adjusted to their limitations they are able perhaps to move upwards in the field and eventually move into field D.
>
> D. It is also possible to find oneself in field D, for example, a person with a handicap or a person with a temporary or long-term illness. From the salutogenic viewpoint, the aim is that even the chronically ill should be able to visit or find themselves in this field as much as possible.

Life is not static. Every human being wanders among the four fields. A philosophical reflection about the importance of health and ill-health can also give us arguments for asserting that illness, strain and burden have a necessary place in human life. Hans Selye[42] maintained that an imposed stress which alternates with rest and recovery is something which strengthens the human organism. Viktor Frankl suggested that all suffering is not meaningless, but can only be understood when one has emerged from it.

Summary

There are thus always two starting points- ill-health or health.

Salutogenesis means the origin of health and implies that instead we ask how it is that in spite of everything health is preserved or even improves.

Pathogenesis means that we try to find out what causes ill-health and illness.

Explaining what health is

A *dichotomous* viewpoint of health implies that we try as exactly as possible to diagnose whether a person is *either* fit *or* ill.

From a *continuum viewpoint*, health becomes something which every person enjoys, to a greater or lesser degree, all the time. At each moment of time, we find ourselves at some point on the axis between the poles of complete ill-health and complete health. By adopting this viewpoint, health promotion aims at creating the preconditions for moving to greater health.

The character of the two perspectives are summarized in fig 3.3

Salutogenesis	• A continuum viewpoint • Knowledge about health determinants • Focus on the whole human being and the conditions which mould our lifes (holistic approach) • Strategy: promoting health
Pathogenesis	• A dichotomous viewpoint • Knowledge about causes of illness, it's symptoms and treatment. • Focus on the sick part (atomistic approach) • Strategy: curing and preventing

FIG. 3.3 *Two ideas: Salutogenesis and Pathogenesis*

The *two dimensional model of health* is a graphic model which is used to illustrate what health is. In it we have both aspects of *health* and *ill-health*. We look upon the state of health as a *continuum*. The two dimensional model also embodies both the physiological or clinical dimension of health and the *psychological* aspect of how health is *experienced*. In the work of health promotion, it is particularly interesting to think about how we define health. How do we recognise health? It is important to include both the physical and psychological aspects of health. The way we describe health

Chapter 3

becomes a criterion for comparing, for example, various factors in an occupational situation. What is it in our work which helps people to feel well and enjoy good health?

4. Health promotion- a long tradition

In this chapter we shall try to understand what health promotion is and its potential role, and how it can be developed by examining its background. We begin with the following questions:

- Where do the ideas come from and on what values are they based?
- Where do the concepts come from and have they a historical and scientific basis?

From a knowledge of the origin of the ideas and of the importance of the concepts, we can obtain a conceptual and theoretical starting point which helps us to understand the nature of health promotion. The term health promotion is an English term which has been adopted in Swedish but with some differences in meaning. There is no unique agreed definition of this concept in the scientific literature. One may feel that the word promotion would mean a purely salutogenic approach but in fact it is often used to cover health work which contains both preventive and salutogenic measures.

There are reasons for isolating and restricting the meaning of the concept. It is a matter of the development of knowledge and evaluation within a new field. In order to be able to study and document a field of activity, we have to be clear about where to draw its boundaries and about how concepts are to be defined. The prevention of illness and health promotion are conceptually two quite distinct starting points. The prevention of illness begins with risk factors for illness (pathogenic factors) whereas health promotion

Chapter 4

begins with health factors which promote health (salutogenic factors). When we use these respective approaches to arrive at knowledge, we must ensure that these concepts are clear and are not confused with one another. We must be aware of the idea or approach that we have adopted so that we know how to evaluate the results we obtain and also how we describe what we are doing and the change which occurs. It is also the case that in evaluating the respective approaches, different methods are used and different questions are asked. It is perhaps more natural that a person who has expertise in social medicine evaluates measures intended to prevent illness. On the other hand, salutogenic measures should be evaluated by someone from the social and behavioural sciences.

Isolating the concepts is also valuable when knowledge via educational instruction or training, is to be translated into practice. We need to develop tools and methods which are tailored to the respective approach adopted in our health work. It is a question of finding and using the right knowledge for the purpose in hand. We must ask ourselves: do we need knowledge which is based on medicine or on behavioural science?

At the same time, it is important to stress the importance of the fact that various ideas, disciplines and methods can be on good terms with one another and co-operate. In everyday situations in a particular workplace, it is not access to certain competence or tools, which should govern the contents of the health program. It has to be the specific conditions which apply to the particular workplace and the people employed there which decides "who, what, and how" in planning what is to be done. In order to achieve success, it is necessary to have both an overall view of things as a whole and co-operation. This co-operation must occur both between the potential clients and the health workers and between the various professions who are concerned with prevention, rehabilitation and a salutogenic approach in their work.

Health promotion- a long tradition

Note that the distinction we make between prevention and the salutogenic approach is not aimed at creating a conflict between these approaches. Antonovsky[43] emphasises that we must look upon them as complementary. They can and ought to be carried on in parallel and the particular situation decides if and how they can contribute to the goals that have been set and to helping to achieve the prerequisites for improved health.

Health *promotion* is a concept which is increasingly used in Sweden. We encounter it and similar expressions in the daily press, in books like the present one and in official documents. Since the 1990s, it has become increasingly common to speak of '*promoting health*'.

Initially the word's English origin (the corresponding Swedish expression is *främjande*) made it seem somewhat foreign. The first three year academic course specialising in health promotion was started at the School for Health Sciences in Vänersborg in 1994[44]. A Master's degree in Health Promotion was inaugurated at Uppsala University in the autumn of 2002.

Health promotion has become an established concept. New projects are being started all the time designed to give concrete expression to the concept. There are projects conducted within the private sector and in the public sector at both municipal and county level, which are aimed at improving health. New private companies are launched by entrepreneurs inspired by the new ideas who see business opportunities within the field of health promotion. Collectively these things show that there is a desire to find new ways of improving people's health.

The development of new work systems takes time. New trends begin often as ideas in someone's brain. They have to be formulated, developed and tested. In order to find the best solutions, the process has to be subject to critical examination. This forces thinkers and

Chapter 4

entrepreneurs to constant refinement of their theories and tools. The critical examination is carried out not only by objective scientists and perhaps journalists. When new ideas arise in a field, it is often the established actors with their established methods who become the sharpest critics and often the chief opponents of the new approach. As far as the idea of health promotion is concerned, it is the medical professions which have played that role. When first launched in Sweden in the 1970s, ideas that health promotion activities could have a complementary role in health work, were treated as scientifically dubious, unstructured and viewed as unwelcome competitors for funds.

One example of changes in health work is Occupational Health and Safety (OHS), where competence in occupational medicine has always played a strong and dominant role. Changes in working life have, however, entailed an ever increasing need for psychological and pedagogical expertise. The work of OHS has correspondingly expanded and new professional groups have played an increasing role in this expansion. New team-members such as behavioural scientists, health educators and wellness and fitness personnel lacking a medical background have found their skills brought into question by more established professions such as physicians, nurses, physiotherapists and engineers.

The encounter between these different professions and their corresponding cultures gives rise to difficulties in the way things are perceived and how language is used. A new common culture does not spring up unaided: it requires openness, dialogue and it takes time. Within the sphere of company health, where this work has been allocated a role and has been consciously pursued, scepticism and opposition has been replaced by teamwork. As a result, it has brought benefit to the clients.

The historical background

The concept of health promotion and interest in this approach to health work has a historical background. It is useful to sketch this in order to understand both how the term came about and the meaning attached to it.

We shall choose certain key events and documents which have played an important role in the development of health promotion up to the present day. These examples form only a fraction of the material which has been published in the domain of health promotion. However, they can be seen as important and, given the aim of the book, suffice to describe the conceptual and theoretical foundation upon which health promotion rests.

Apart from the material presented here, many research conferences have taken place and many network meetings have been organised under the program title of health promotion. At these meetings, new ideas have been put forward and tested. Researchers from various countries have given an account of their experiences and conclusions derived from their work on health promotion. It is impossible to summarise all this in the scope of the present book.

As our starting point, we shall begin with the following documents which are presented in historical order:

- WHO's constitution 1948
- the LaLonde report 1974
- The WHO conference in Ottawa 1986
- the European Foundation Workshop in Dublin 1992
- The Luxembourg declaration 1997

Chapter 4

WHO established in 1948

The historical starting point could well have been set several hundred or indeed several thousand years back in time. This we have partly done in chapter two. Here, we shall take as our starting point one which will lead to a contemporary description of what health promotion is and how it can be applied. The idea is that finally we shall come upon work systems which fit in with modern workplaces in modern organisations. The most relevant knowledge for this task is therefore not so very old. The majority of authors, who describe health promotion and its conceptual background, take as their starting point the creation of WHO in 1948. The documents which originate from this event are the first to contain a more or less radically new view of the concept of health. What is new is the acceptance of a holistic concept of health as something which is more than merely the absence of illness. Health is described as a state of complete physical, mental and social well-being.

In the WHO constitution, we find the following[45]:

> Health is a state of complete physical, mental and social well-being and not merely the absence of disease or infirmity.
>
> The enjoyment of the highest attainable standard of health is one of the fundamental rights of every human being without distinction of race, religion, political belief, economic or social condition.
>
> The health of all peoples is fundamental to the attainment of peace and security and is dependent upon the fullest co-operation of individuals and States.
>
> The achievement of any State in the promotion and protection of health is of value to all.

Health promotion- a long tradition

Today we consider this definition of health unnecessarily Utopian. However, it has had considerable importance for the development of health promotion because of its stress on the fact that health is more than simply a physiological and medical issue. WHO also includes the mental and social dimensions as part of their definition of health. Thus it becomes important to ask people about how they themselves view their health and well-being. We can combine the description of health with both physiological and psychological measurements as in the two dimensional model of health (see page 59).

Through the years, the social dimension has played an important role in the work of WHO. Health is dependent both upon social factors such as economic opportunities, knowledge/education and equality. The goal is health for everyone and the concept of *empowerment* is often used for the political and social measures designed to give people possibilities and assets for dealing with their health. The work aims at promoting health and preventing illness. When WHO included well-being as part of its definition of health, this has had the positive effect that individuals themselves are involved in deciding about their state of health. According to certain researchers, there can be a danger in completely transferring the decision to the individual.[46] What is the cause or basis of the well-being experienced by a person? People experience well-being on the basis of their own criteria and experience, which e.g. an addict may do in very unhealthy circumstances. With some support, however, the person would be able to improve both their physical status and the well-being they experience.

LaLonde report 1974

A later document advocating a health promotion approach is Marc Lalonde's report[47] from 1974 where he presents his "health field concept" which attempts to describe the determining factors of

Chapter 4

health from a holistic perspective. There are, according to Lalonde, four main areas which contain factors which influence health.

- human biology
- environment
- life style
- health care organisation

Lalonde set out to find alternative approaches to Canadian health work in order to put an end to the spiralling costs of health care. Among a number of proposed solutions, he formulated a *Health Promotion Strategy* and described it as follows:

> A Health Promotion Strategy aimed at informing, influencing and assisting both individuals and organisations so that they will accept more responsibility and be more active in matters affecting mental and physical health (p. 66).

Lalonde also presented a detailed list[48] of concrete programs and measures which a health promotion strategy should contain. Here are some examples from the list:

1. The development for the general public of educational programs on nutrition.
2. The enlistment of the help of the food and restaurant industries in making known the calorific value and nutritional content of the food they sell.
8. Encouragement among employers of programs designed to ease the transition from employment to retirement.
9. Reinforcement of successful programs for making life more interesting for the aged.

Health promotion- a long tradition

10. Promotion and co-ordination of school and adult health education programs, particularly by health professionals and school teachers.

12. Continued and expanded marketing programs for promoting and increasing physical activity by Canadians.

13. Enlistment of the support of the educational system in increasing opportunities for mass physical recreation in primary and secondary schools, in community colleges and in universities.

14. Promotion of the development of simple intensive-use facilities for more physical recreation including fitness trails, nature trails, ski trails, facilities for court games, playing fields, bicycle paths and skating rinks.

15. Continued pressing for full community use of present outdoor and indoor recreation facilities, including gymnasia, pools, playing fields and arenas.

16. Continued and reinforced support for sports programs involving large numbers of Canadians.

18. Extension of present support for special programs of physical activity for native peoples, the handicapped, the aged and the economically deprived.

19. Enlistment of the support of women's movements in getting more mass physical recreation programs for females, including school children, young adults, housewives and employees.

20. Enlistment of the support of employers of sedentary workers in the establishment of employee exercise programs.

21. Enlistment of the support of trade unions representing sedentary workers in obtaining employee exercise programs.

22. Increase in the awareness of health professionals of factors affecting physical fitness.

23. Completion of the development of a home fitness test to enable Canadians to evaluate their fitness level.

These are some of the points in Lalonde's list. It is a typical example of the 1970s' focus on lifestyle issues as important ways of promoting health. Measures such as education, information, efforts to inculcate a better diet, more physical activity and various social activities dominate the list. The recommendation having the widest coverage was that emphasising physical activities. This was clearly an important area; In Sweden at that time, prominence was being given to just such physical activities within the framework of Fitness and Wellness programs. The National Board of Health and Welfare agitated about the need for preventive measures against ill-health, most importantly in the form of physical activities. The athletics and sports associations were already perceived as important actors in the campaign to promote public health and in the government's budget, extra funds were made available in order to create facilities which would allow more people to be physically active. A particular investment in sports for the masses making use of the Swedish Sports Confederation and various special sports associations was one of the results of this in the 1970s.

Ottawa Charter 1986

The perhaps clearest and most important occasion for the development of health promotion, as we describe it today, is the "First International Conference on Health Promotion" which WHO held in Ottawa in November 1986. It was at this meeting that the

Ottawa Charter for Health Promotion[49] was set forth. This text has a clear ideological foundation with its emphasis on *participation* and *equality* as prerequisites for creating health. In this document, we find also the basis for the formulations and several of the principles which are central to the literature which today describes health promotion. In the Ottawa Charter, health promotion is described as "the *process* of enabling people *to increase control* over, and to improve, their health." (my italics). Furthermore, health is seen as a resource in the case of the individual and is dependent both on a wise health policy and an environment that is supportive and stimulating.

In the Ottawa Charter, there are a number of central themes which deserve to be examined more closely.

Focus on the promotion of health as a resource

Health cannot simply be defined as the absence of illness: it has to be seen as a human resource in daily life. The goal is not good health as a value in itself: we seek good health because it allows us to attain our goals and wishes and improve the quality of our lives in other ways. The Ottawa charter puts it like this:

> Health is, therefore, seen as a resource for everyday life, not the objective of living. Health is a positive concept emphasising social and personal resources, as well as physical capacities. (Page 5)

And later on the same page:

> Good health is a major resource for social, economic and personal development and an important dimension of quality of life.

Chapter 4

By thus describing health as a resource for social, economic and human development, WHO made clear that one wanted to give a health promotional direction to health work. The prerequisites and prospects for health cannot be ensured by the health sector alone. More importantly, health promotion demands co-ordinated action by all concerned parties in society: the government, the administrative and executive bodies, the media, and the traditional social and health sectors. At the local level, both individuals, companies and municipal authorities must be involved in the work. In this way, health promotion becomes a broad strategy for public health, able to deal with political, economic, social, cultural, environmental, life-style and other factors in such a way as to yield effects on health which are as beneficial as possible. In the Ottawa Charter, health promotion is described as a new dimension of strategic thinking about health and thus becomes of central importance in the policy of *health for all* which was launched earlier by WHO at its conference in Alma Ata in 1978.

Participation

According to the Ottawa Charter, health promotion strives to reduce the differences in the state of health among human beings and to provide everyone with equal opportunities to attain good health. For this to be the case, people have to participate and acquire greater possibilities for influencing the factors which are important for health. It is a question of making things possible by imparting knowledge and the power to influence. The same possibilities are to made available to everyone, whether they be women or men, poor or rich. The term empowerment has often been used to describe this aspect of increased autonomy and everyday power or authority which has to be available in order for people to have or take control over the preconditions or circumstances determining their health. As the Ottawa Charter[50] puts it:

Health promotion- a long tradition

Health promotion is the process of enabling people to increase control over, and to improve, their health....People cannot achieve their fullest health potential unless they are able to take control of those things which determine their health (p.5)....At the heart of this process is the empowerment of communities- their ownership and control of their own endeavours and destinies (p6)...Health is created by caring for oneself and others, by being able to take decisions and have control over ones life circumstances. (p7)

A society which allows people to have influence on their health, must carry out a policy which leads to its citizens having knowledge about their own life situation and also influence over it. Good public health presupposes also an equal distribution of the prerequisites for health among the population. In particular, the weaker groups in society must receive various kinds of support to allow them to avail themselves of opportunities presented by their health and to make healthy choices when it comes to consumption and life style. Just in this respect, we know that healthwise there is an inequitable distribution between different groups in the population. Educational level, income, residential area etc are correlated with state of health. People who are well off in the economic and social sense, have also, on average, better health. This observation is also significant for the distribution of health in working life. It is easy to come up with principles for participation and employee influence intended to apply to all workplaces. In practice, however, difficulties arise even here and not simply because there are physical differences in environment between different workplaces. Company values and culture, differences between various trades or professions, educational level and traditions of participation and influence vary between work groups and between organizations.

Chapter 4
The setting - a situational starting point

Health is influenced and created where people are employed It is both people themselves, their personal preconditions and their environment which affect their health. Health promotion must therefore cover both people themselves and the situation of which they form part. Physical and social environment, as well as people's thoughts and actions form part of this situation. By anchoring health work by thinking in terms of *setting*, it becomes possible to carry out a health promotion strategy in a more meaningful way. The setting is a concrete situation where there are people and shared conditions and where one can work together to influence these conditions and improve the preconditions for health. WHO has taken the initiative to focus on settings such as school (in Sweden, called the health promoting school), the local community, hospitals (the health promoting hospital) and the workplace.

Heath promotion aims at creating supportive environments where health can be preserved and improved. The Ottawa Charter puts it like this

> Health promotion works through concrete and effective community action in setting priorities, making decisions, planning strategies and implementing them to achieve better health (p. 6).
>
> Health is created and lived by people within the settings of their everyday life; where they learn, work, play and love. (p.7)

A historical analysis of views of health promotion work shows that the approach has varied. The ideas and literature which was produced during the 1970s was geared to the individual and people's life style as an important domain which could be influenced and changed. Health becomes what I myself make of it by my own way of living and the extent to which I care about what is good for body and soul. In the Ottawa Charter and in subsequent documents, we find throughout the 1980s a greater emphasis on the significance of the surrounding environment and the social context. During this period, more and more has been written about the unequal distribution of health in society. Research[51] has also been able to demonstrate clearly that socio-economic factors have a strong connection with people's health. As one sometimes puts it: "it is better to be rich and healthy than poor and ill".

The Ottawa Charter, with its global perspective, lays great emphasis on the importance of the total living environment when it comes to health:

Prerequisites for Health

The fundamental conditions and resources for health are:

- peace
- shelter
- education
- food
- income
- a stable eco-system
- sustainable resources
- social justice and equity

We regard the majority of the above prerequisites or preconditions for life as necessary and perhaps even as self-evident for some of us. Possibly it is other environmental factors which we in Europe, for example, have insufficient control over. Everywhere the living environment has its own unique properties and effects on human beings.

Chapter 4

People's personal prerequisites and the environment they live in, are two starting points or perspectives which are important for health. To these two, a third deserves to be appended. It is presumably neither the qualities of human beings nor of their environment which have the greatest importance, but rather the encounter or interaction between human beings and the setting they find themselves in. According to Antonovsky[52], who describes in his research how people, despite the shortcomings in their circumstances, nevertheless preserve their health. What is decisive is the quality of life which people experience in the setting they find themselves in. It is perhaps just this which is the really interesting thing. We live in a situation or setting and despite our lives being as they are, some of us can still enjoy a quality of life and preserve our health. How is this possible? What are the health factors which, no matter the time and situation we find ourselves, allows us to cope, feel well and preserve our health despite the fact that we live as we do?

The word *process* recurs in the Ottawa Charter and in many other documents on health promotion. It means that health is not something which is swiftly and easily fixed. Instead, it is a process which takes time and which sets out to give the people concerned, power and possibilities. Processes in social systems are unpredictable and difficult to control. It is a question of living with things, communicating, interpreting and trying to understand what is happening, and how one can influence the system.

WHO set up a visionary target of perfect physical, mental and social well-being. For most people, such an aim is unrealistically utopian and perhaps not even desirable. Good health means so many different things for different people in different parts of the world. A vision of perfect health such as that of the WHO definition can nevertheless serve as a pointer when health work is being planned or carried out. The vision should then be followed up by strategies and methods for the analysis of needs, decisions, planning and the improvement of the conditions for health. The process way of thinking and working aims at a long-term and systematic application of the health promotion project, a prerequisite for this type of health work.

The concept of process also represents a determination to see that the health issue is always part of the daily agenda in all those circumstances where decisions, possibly affecting people's health, are made. A process oriented way of working implies that constant check-ups and evaluation provide new knowledge which then interacts as feedback with the health promotion process and means that it is continuously improved. The process way of thinking has given rise to many different working models and systems which often follow a PDCA (Plan, Do, Check, Act) form. Generally the process approach is about using resources in an effective manner with the right measures for the right people at the right time. The process derives its energy from the fact that there is a dialogue between the people involved and different organisational levels. The dialogue gives concrete form to participation and makes use of people's ideas and knowledge. This in turn is an important prerequisite for involvement and the assumption of responsibility, which are key values in a properly functioning health promotion process. The application of a process-oriented way of working varies according to the setting we are working in. The workplace and organisation are well suited to this approach and we shall return to this in chapter twelve.

Health education as a part of health promotion

In the international tradition of health promotion following the Ottawa Charter, there has been great emphasis on informing people about health and the preconditions for health. The idea is that if people are informed about what is respectively good or bad

Chapter 4

for their health, they will be in a better position to make their own considered judgement about how to attain good health. As the Ottawa Charter states[53]:

> Health Promotion supports personal and social development through providing information, education for health, and enhancing life skills. By so doing, it increases the options available to people to exercise more control over their own health and over their environments, and to make choices conducive to health.(p.7)

This aspect of health work has various names. In English-speaking countries it is called *health education*. Its ultimate aim is that through the provision of information and knowledge, we shall help to achieve a more healthy lifestyle among the population. Health education entails both public information campaigns directed at various population groups or risk groups, as well as educational activities, e.g. in the school or workplace. We can target a group or individuals one at a time.

Health education, as a preventive measure, can consist of specially designed campaigns to prevent certain illnesses or groups of illnesses. One recurring activity through the years has been anti-smoking campaigns where one begins with smoking's role as a risk factor in cardiovascular illnesses and cancer. When people are informed about this, fewer of them choose to begin smoking, assuming of course, that the campaign is successful.

It is also possible to look upon health education as a salutogenic health promotional measure where information and knowledge is sent out as a general measure in order to make groups or individuals better informed and more aware about what is beneficial to health. In this way, we can give people the opportunity to make healthy choices and to attain a healthy lifestyle, and provide the motivation for doing so. In the case of school pupils in the age band 12-13, for example, it is a salutogenic health promotional measure to reinforce their confidence so that they have the courage to make independent decisions. This then becomes a resource which may be expected to lead to more of them saying "no" to alcohol and tobacco, which in turn is good, both for the individual and the health of the population at large. A prerequisite is that this initiative is combined with knowledge about the risks associated with the use of alcohol and tobacco.

The foregoing example is one of several where we can say that in practice the preventive and health promotion approaches work together co-operatively. The starting point can be an increasing problem of ill health which gives a reason for picking out a target group and creating resources. Information about the injurious effects of tobacco is presented in order to prevent tobacco-related illnesses: it is thus a preventive strategy. The business of strengthening children's self confidence is generally good for their health and as such, is part of a health promotion strategy.

The words we choose to describe what we are doing, can seem unimportant: the main thing is the results we hope to achieve with our health work. It is the ability of the health educationist and the other actors involved to make use of several strategies and, depending on the groups involved and the objective, to design an effective program which will decide what results are achieved. As mentioned earlier, the choice of strategy is important when it comes to evaluating the effects of the measures.

European network for health promotion in working life

WHO has continued its work on the basis of its global perspective. Since the Ottawa Charter, working life health issues have continually been on the agenda. Among other things, a conference was held in Jakarta in 1997[54] which emphasised the need of evolving strategies for health which take account of the bigger picture. Health work with the workplace as setting must take account of physical, emotional, psychosocial, organisational and economic factors. WHO developed a new global initiative called WHO;s global Healthy Work Approach (HWA). This would become an inspiration and brought together various interested parties who wanted to pursue a health promotion strategy in concert with *Occupational Health and Safety*, *Human Resource Management* and *sustainable development*.

In now going on to present our account of the origin of health promotion with a focus on working life, we shall confine ourselves to a European Union perspective. As has been mentioned above, there are still many initiatives concerning health promotion to be found within WHO and in individual countries around the world. From a Swedish point of view, however, it is contacts and co-operation within Europe which has meant most when we are explaining what we mean by health promotion as a health strategy in the workplace.

European co-operation has throughout been based on the principles which were set forth in the Ottawa Charter of 1986. It has focussed on adapting and applying these principles and developing guidelines and criteria for the practical application of what has come to be called Workplace Health Promotion (WHP).

European co-operation is rooted in the establishment of the European Union (EU) in 1991. EU laid stress on the importance of working for the safety and health of people in working life. One finds for example the following statement: "A healthy, motivated and well-qualified workforce is fundamental to the future social and economic wellbeing of the European Union".[55] Since 1975, there has been a European Foundation for the Improvement of Living and

Health promotion- a long tradition

Working Conditions[56] which has actively worked with programs for designing and planning better living and working conditions for the inhabitants of the member countries. This organisation was given the task and carried out between 1989 and 1992 an investigation[57] of 1400 companies and organisations in 8 EU member countries. The aim was to examine the health work being carried out in European workplaces. In this study, it was found that most of the health measures fell within the area of occupational health and safety (OHS). In the material as a whole, there are only a few examples of measures for increased health such as organisational issues, changes in working hours and personal development.

1992 was the EU European Year of Health and Safety at Work. An international conference was held in Dublin from the second to the fourth of December with participants from companies, trade unions, authorities and research. The results of the study of 1400 companies were discussed and a policy document[58] was drawn up, describing important steps in the future development of workplace health promotion.

In developing Workplace Health Promotion, four areas were singled out for attention:

A. Marketing of and Incentives for Workplace Health Promotion

B. Organisational Change for Health

C. Professional Development for Health at Work.

D. Implementation - Strategies, Instruments and Methods.

In each of these areas, several difficulties and issues which needed to be resolved, were described. It is worth thinking about the extent to which these problems still lack satisfactory solutions.

The conclusions of the Dublin conference were afterwards summed up as follows:

Chapter 4

- WHP is a new concept where there is a lack of agreement about how it is to be defined and explained. There is a need to produce a more agreed and unambiguous description of what WHP is, as far as ideas, theory and specific content are concerned. Research and discussion between the actors involved in health work in the workplace must transcend disciplinary and professional boundaries.

- Health in the workplace is no longer a purely medical issue but must cover the work situation and organisation as a whole. It is yet to be tested how far WHP can be incorporated within traditional organisational systems and processes for change. The health issue must be placed on the agenda of decision-making bodies in the organisation. Both management and all the members of the workforce must be involved.

- There is no clear picture of what skills are needed or what form education in WHP should take. (Only certain countries, among them Sweden, had at this point of time educational courses which addressed the subjects and areas which relate to health promotion and WHP in particular). Some are concerned with the education of existing professions in the personnel area and in company occupational health care. It was considered that new professions with academic education in the field would be needed. One problem was that most of the educational courses had a medical starting point. Literature and educational courses in WHP drew their knowledge from the social and behavioural fields to a much lesser extent.

Health promotion- a long tradition

- One of the most important explanations why WHP had not succeeded, was the lack of adequate approaches and methods designed to implement the WHP initiative in workplaces. A development process aimed at health in the workplace requires tools and work routines for analysis of the current situation, and planning which involves many activities for work at the individual, group and organisational level, as well as models for evaluation. In many workplaces, the corporate culture lacked a tradition of participation and failed to allow the individual employee to exercise influence

The Luxembourg Declaration and a new network

After the Dublin Conference, the EU Commission in 1996 took the initiative to setting up the European Network for Workplace Health Promotion (ENWHP)[59]. This network includes the EU member states together with Norway, Island and Liechtenstein. The network participants are national institutes, authorities and universities with knowledge and interest in occupational and public health. A secretariat has been set up at the European Information Centre at BKK[60] in Essen in Germany. The aim of the network is to work towards the vision of "healthy people in healthy organisations". The work is intended to help and support good practical examples of WHP. Among the documents which have been drawn up by ENWHP, the so-called Luxembourg Declaration is extremely valuable in describing what WHP is, and in proposing guidelines for how health promotion activities should be carried out in the workplace.

The Luxembourg Declaration[61] based on the meeting of the network in November 1997 begins by stating that WHP is an

Chapter 4

aspiration shared by employers, employees and society, to improve the health and well-being of people in working life. This can be achieved by combining measures to improve the organisation and the working environment, by increasing the opportunities for employees to participate, and by encouraging and supporting the personal development of employees.

The document stresses that traditional preventive work has been successful in reducing the number of accidents and occupational diseases:

> Traditional OHS[62] has significantly improved health in the workplace by reducing accidents and preventing occupational diseases. However, it has become obvious that OHS alone cannot address the wide range of issues mentioned above (p.2)

Today, however, it is completely clear that preventive work does not suffice in the battle against ill-health or in the task of creating conditions for increased health in working life. Since the Luxembourg meeting, it has become increasingly clear that several strategies are required.

WHP can contribute not only to lower figures of absence due to illness but also to increased productivity. WHP can be a successful modern business strategy which reduces ill-health and increases productivity by means of a workforce with good health, high motivation and a good working climate.

> WHP is a modern corporate strategy which aims at preventing ill-health at work (including work-related diseases, accidents, injuries, occupational diseases and stress) and enhancing health-promoting potentials and well-being in the workforce (p.2)

Health promotion- a long tradition

According to the Luxembourg Declaration, WHP can contribute to health in the workplace by means of a wide range of factors:

- management principles and methods which recognise that workforce is a necessary factor in the success of the organisation, not merely a cost factor
- an organisational culture which relies on the participation and encourages the involvement of all the workforce
- organisational principles which provide the employees with an appropriate balance between job demands and control with regard to work, skills and social support.
- a personnel policy which actively incorporates health promotion issues
- integrated occupational health care.

In the Luxembourg Declaration, ENWHP provides the following guidelines for health promotion work:

> WHP can reach the aim "healthy people in healthy organisations" if it is oriented along the following guidelines
>
> 1. All staff have to be involved (participation). (p.2)

This first guideline is central and not open to discussion in the majority of documents dealing with health promotion. A health promotion approach is based upon the participation, involvement and responsibility of all the people belonging to the setting we consider working in. In a workplace, management leads the way and creates the preconditions. However, it is the employees themselves who have the greatest knowledge about how their workplace actually functions and about what they can and wish to do in order to develop conditions conducive to better health.

Chapter 4

    2. WHP has to be integrated in all important decisions and in all areas of organisations (integration).

Health work must be incorporated in the regular work and decision-making. With the workplace as a setting for health work, we find ourselves entering a technical and social system which involves a host of interested parties and subprocesses. The larger the organisation, the more complex it is. Thinking in terms of setting provides a useful introduction to the everyday situation of the work group together with the relevant manager. It is impossible, however, to carry out health work based on WHP which is confined to a particular workplace, particularly if it is part of a larger organisation. Health work must deal with at least three system levels associated with the workplace, namely the organisation, the work group and the individual.

    3. All measures and programs have to be oriented to a problem-solving cycle: needs analysis, setting priorities, planning, implementation, continuous control and evaluation (project management) (p3).

What the Luxembourg Declaration calls a problem-solving cycle is a way of applying WHO's idea, as expressed in the Ottawa Charter, that health promotion is a *process* for improved health. It is a question of a long term project rather than quick short term measures. The idea of process is based on the insight that there is no single tailor-made solution for improved health which works in *all* workplaces. Every situation has its own particular preconditions which must be taken into account when designing health promotion work. The process is carried out with the participation of many people in the analysis of the current situation – priorities- planning-

Health promotion- a long tradition

implementation- checking up and evaluating. In order to achieve structure and efficiency, a project approach is often employed. (In this connection, there is no conflict between project and process, rather co-operation).

> 4. WHP includes individual-directed and environment-directed measures from various fields. It combines the strategy of risk reduction with the strategy of the development of protection factors and health potentials (comprehensiveness) (p.3).

In this fourth guideline, the Luxembourg Declaration provides a further meaning of WHP. Whereas the Ottawa Charter is more clearly directed at a salutogenic approach, emphasising prerequisites for health rather than risks for illness, ENWHP in the above passage allows Workplace Health Promotion to include both preventive and salutogenic strategies. According to this, WHP must cover both risk reduction, safety work and measures that actively promote health (health potential). In this way, WHP become an umbrella term for both the preventive and promotional approaches.

In its introduction, the Luxembourg Declaration comes out clearly in favour of the latter when it says that:

> Workplace Health Promotion (WHP) is the combined efforts of employers, employees and society to *improve* the health and well-being of people at work. This can be achieved through a combination of:

- improving the work organisation and the working environment
- promoting active participation
- encouraging personal development (p.1)

Chapter 4

This text is closely linked to WHO's intentions and expresses clearly what WHP means by a salutogenic approach. A first important requirement is co-operation between the parties involved. WHP involves measures to develop both the workplace and people. This work is based on active participation.

There are various reasons behind the desire to make WHP on the one hand an umbrella term and on the other hand a concept which is restricted to salutogenic health work. The umbrella term can be helpful in countries which completely lack a tradition of health work in occupational life. In such places, there is a need for the development and introduction of strategies of all kinds. On the other hand, in Swedish working life for example, we have a long and well developed tradition in the preventive domain, thanks to occupational health and safety work for some fifty years. For that reason, we have more to gain, I would maintain, from a clearer and more strictly defined definition of what it means to treat health promotion as a third strategy, a salutogenic one. There is a need for a clear conceptual starting point and direction in order to be able effectively to develop health promotion as an area of knowledge and methods which are adapted to it.

After the conference in Luxembourg, the network's activities have continued and new policy documents have been drawn up.

The Cardiff Memorandum[63] from the 1998 network meeting takes up the need for special investment in small and medium sized enterprises[64]. In these enterprises there are fewer possibilities for implementing WHP. This is something that is familiar in Sweden where we are aware that small enterprises have fewer economic and personnel resources to invest in health work. In addition, fewer workplaces belonging to this group, make provision for company health and safety measures.

Health promotion - a long tradition

In the document, ENWHP clearly supports an increased investment in small and medium sized enterprises and wishes to develop WHP models and strategies which are adapted to their needs.

The *Lisbon Statement of 2001 on Workplace Health in SME:s*[65] once again discusses the initiative with respect to small and medium sized enterprises. In this document, stress is laid on the great importance of smaller companies for the European economy. The innovative capacity and economic results of such enterprises very much depend on the health and well-being of their personnel. According to the document, these small companies are in themselves a health resource. In them, one often finds a family atmosphere, nearness, a good overview of what is going on and a possibility for increased participation and responsibility for the individual employee. Health work should go hand in hand with the firm's business and organisational development. Work aimed at improving health in these enterprises implies that the management must incorporate health issues in daily decision-making and involve their employees in the work of improvement which concerns both the working organisation, the atmosphere at work and performance. In order to get things rolling, ENWHP wants every nation to engage their respective authorities and other concerned organisations linked to working life and health, so that it will be possible to disseminate ideas and knowledge about strategies for better health in small and medium sized enterprises.

The *Barcelona Declaration on Developing Good Workplace Health Practice in Europe*[66] is the product of the third European conference on Workplace Health Promotion which was held in the summer of 2002.

Chapter 4

This document does not add much that is new. The importance of the workplace for both the individual's own "health practice" and public health from a national perspective is underlined.

> "individual health practices are shaped by our workplace cultures and values"

> No public health without good workplace health".

In the document, it is stated that despite clear evidence of its value, there are still only a few European enterprises which have begun to implement clear policies and strategies in regard to WHP. In many countries, knowledge and methods for analysis of the current situation and for the implementation of WHP are lacking. In order to find a remedy for this, ENWHP wants to create a "toolbox" for that purpose. Otherwise, the Barcelona Declaration, like the Lisbon Statement of 2001, points out that every nation must create its own agenda for *workplace health improvements* (p.3). Every nation has to draw up a list of methods and instruments for WHP and has to engage in the analysis of health work's economic consequences for enterprises and organisations. An account of this work was presented at a conference in Dublin in June 2004.

By way of summary, we can say that there is an international interest in describing all health work as health promoting activity. The reasons for this are perhaps to be found in the strong tradition and predominance which the disciplines of medicine and social medicine still have. All this knowledge and all its representatives must continue to have a central role in health work. But it is sometimes difficult for new and initially somewhat unconventional ideas to make themselves heard and win respect. When the idea of salutogenesis came on the scene, some of the medical establishment reacted by saying: "Oh that! We have always thought that way and acted correspondingly ; there's nothing new about it!". Or else they maintained it was *to some extent* new, an idea and approach which might conceivably have a role to play alongside the traditional medical one. When we then set out to acquire knowledge about what salutogenic health promotion means, the greater part of research -as well as the prior right to interpret the results- ends up in Schools of Medicine. It is after all their representatives who traditionally have explained for politicians and the general public what health is and how one should deal with it. The fact that the discussion has usually taken ill-health as its starting point, is simply considered an odd detail in this context.

There are today increasingly more people[67] who hold that medical theory and science do not suffice in health promotion work. Nor is it sufficient to complement that with health psychology, as has occurred in recent years. These two domains, medicine and psychology, still keep us 'inside the skin of human beings'. If the work of health promotion is to be successful, it has to begin with the social and physical situation which people find themselves in. Human individuals are human individuals, but they are also part of a system with different levels each part of the system has its own rules and conditions and these are determined by how the surrounding system levels function and influence things. From a starting point of human beings as individuals, we can progress to system levels further inside e.g. the organic, the cellular and the molecular levels. Knowledge about these parts is to be found in human biology and medicine. If instead we move outwards to other system levels, we can look at human beings e.g. in a group, in a workplace, in a society or in a nation.

The way in which the "inner" system levels function, determines whether health is to be maintained or improved. It is, however, also absolutely necessary to see how the "outer" or larger system levels in

Chapter 4

the surrounding environment function and how they affect people's health.

What form should health promotion take in practice? There are many theories and disciplines which can contribute knowledge which is useful for health work. The aim of the present book is to draw attention to knowledge which can be of assistance in developing salutogenic health promotion. There is thus reason to pose the following question: which theories and theoretical knowledge can be useful in this work? In the next chapter, we shall discuss what we mean by theory and how we should structure all those things which do not deal with concrete activities, but which are nevertheless necessary, in order for us to do - as far as possible - the right things in the right way.

Summary

In order to understand health promotion, we have taken a retrospective look at the history of the subject by examining certain important documents dealing with health promotion. These are

- WHO's Constitution of 1948
- The Lalonde Report of 1974
- The WHO Conference in Ottawa in 1986
- The European Foundation workshop in Dublin in 1992
- The Luxembourg Declaration of 1997

We speak of health promotion as a strategy for developing health in the workplace. The great desire and hope is to develop this strategy and transform it into well functioning processes which really do contribute to better health for people, as well as healthier organisations. To do this, we need a more or less unambiguous, clear and agreed definition of what health promotion is, and how it is carried out in practice. When we look for books, articles and

Health promotion- a long tradition

other textual material to provide us with just such a description, there are masses of things to choose among. The present chapter has taken as its starting point the documents above, which have been published by internationally recognised and important organisations. They stem from the work of researchers, practical people working in the field and representatives of society within the health sector who have brought together their ideas and knowledge. This textual material is theoretical, even if a great deal of it is based on practical experience and research.

In the material from WHO and ENWHP, we fail to find an unambiguous description of health promotion. Many of the documents include both salutogenic health promotion and preventive (pathogenic) work under the heading of health promotion. In this way, the term is used in two distinct ways, partly as a concept referring *exclusively* to salutogenic health promotion and partly as an umbrella term covering *both* promotion *and* prevention.

We note that this ambiguous use of the word occurs in the international literature and we must live with it. However, it can lead to difficulties when we speak about fields of knowledge, different ways of working and different approaches in connection with analysis and evaluation.

It is one thing to know what is good for health i.e. measures promoting health. It is something quite different to know how, in an organisation, we bring about the movement towards the health-pole which is our intention. It is here where the new Swedish approach to health promotion helps us to deal with the particular circumstances in which the health promotion measures are to be implemented. The very word pro-*motion* helps us to see the concept as a strategy for creating a *movement* towards the health-pole. From the central documents of WHO and ENWHP which are cited in this chapter, we shall borrow four criteria which also create a precondition for

Chapter 4

movement and change towards increased health. A more concrete description of the four criteria of salutogenic health promotion will also be given in chapters 9-12, but we mention them already here as a summary of what the literature above gives us. See fig 4.1

Focus on measures **promoting health** (salutogenic idea)	Setting/system thinking (the workplace is the particular context/connection)
Participation (conditions for succeeding)	**Process oriented work** (the approach adopted)

FIG. 4.1 *Four criteria of salutogenic health promotion*

The focus on measures promoting health is directed at the positive, health side of the equation and embodies the idea of salutogenesis and a holistic view of health which means that many factors have to be taken into account.

Another important aspect is *thinking in terms of setting* which means that health promotion places great importance on people's living environment and social nexus. In this connection, a system viewpoint is adopted which means that in a workplace one reflects about the role of individuals, work groups and the organisation in a health perspective.

People's participation is also necessary in order to give legitimacy to what is being done and for involvement, a quality which requires pedagogical insights and takes time to establish. Consequently health promotion work cannot be planned in advance from A to Z. Preconditions can be created, but thereafter the work must be carried out, step-by-step, in co-operation with those most concerned with the issue, namely the workplace employees.

Health promotion - a long tradition

A process-oriented way of working is needed in order to succeed. The workplace and organisation are complex systems which are difficult to predict and steer and the development processes must take their time.

With these four criteria, health promotion is assigned a clear identity of its own which implies that the concept is also differentiated from the composite picture of everything that does or can promote health. An activity which promotes health is thus not the same as health promotion. Health promotion presupposes a salutogenic direction, but also that the other three criteria are present in order that the movement towards improved health takes place. Health promotion implies that the factors or inputs promoting health must be found or placed in a concrete situation or setting.

5. From idea to theory

In order to bring about change and get results, various types of competence are needed. It is possible to describe competence to some extent, but it is something which is also linked to personalities and situations in a way which makes it sometimes difficult to understand what exactly are the qualities which give rise to good results. Putting it simply, one might say that competence is a matter of both knowing and being able to apply that knowledge, two aspects which often go under the names of theory and practice. In the optimal situation, we have access to both aspects simultaneously when we are to carry out some form of activity. If we have only knowledge of methods and practical experience, we run the risk of becoming bogged down in routines and use the "customary" way of solving the problem in hand. We tend to say that practical experience without knowledge is blind in its application. Theoretical knowledge is required for reflecting about and analysing what we are engaged in and what effects it has. On the other hand, theoretical knowledge becomes "vacuous" if it is not used or tested in practical action. By making use of both theoretical and practical knowledge, the work can be planned and executed in a wise and appropriate way relative to the conditions which prevail and the purpose to be achieved.

Taking the above as our starting point, there is reason to examine the theory which is needed and available to design and carry out successful work in health promotion. We also need to find out about the practical application of work in health promotion which has been carried out in different places and which can provide us with useful experiences. The development of new activities often takes place in project form where we plan what we want to do, carry it out and then evaluate the results. A pilot project is the normal way of

Chapter 5

approaching matters when we want to test a new idea or a new way of solving a task. Great importance is attached to evaluation and to the analysis of the execution of the project and the results achieved, in order to decide if the overall results are sufficiently good for the project to assume a more permanent form, and to determine if the experiences derived from the pilot project are worth disseminating to others.

When we transfer ideas and models for health work from one workplace or organisation to another, it is important to distinguish between, on the one hand, knowledge which is general and transferable and, on the other hand, knowledge which is dependent on the particular context and which cannot be transferred or copied. When workplaces and projects are described as exemplifying "best practice" or "good practice" the value of these potential role-models must always be analysed in terms of the particular circumstances which have given rise to them. It is a good idea to search for potential role-models which other workplaces can imitate. It can be thought that a measure or activity which yields good results in one workplace, ought to work in other workplaces. This way of thinking about things can, however, be risky since it is never possible to know in every case the specific conditions which apply to a particular workplace. We thus require some form of what Moldaschl and Brödner[68] call a "reflexive methodology of intervention", a topic we shall return to later in the book.

Theory at different levels

Theory is partially synonymous with the concept of knowledge and we shall start by examining how we can distinguish between the different levels of theory and knowledge which can be used in the work of health promotion. In order to keep close to everyday concepts and ways of speaking, we shall take as our starting point the concepts of theory and practice. Put simply, we can say that

From idea to theory

theories have the answers to questions of why and how a problem is to be solved. Practice, on the other hand, supplies the methods and tools for obtaining a solution.

The concept of theory is in itself wide-ranging and can be employed in various ways. In order to distinguish between the different forms of theory, we shall divide them into groups, depending on their level of abstraction. The concept of theory can thus be more closely linked to practice and describe a more concrete level of knowledge. Proceeding from this level, we can then make the theory more abstract, moving, as it were, up a ladder to higher theoretical levels, so-called levels of abstraction. The further we distance ourselves from practical action, the more philosophical the theoretical level becomes. We can also move in the opposite direction by beginning with an idea or thought of a more philosophical nature which we then exemplify in terms of knowledge which is progressively more practical in character. The idea or philosophy then leads to practical action which is carried out with the help of knowledge relevant to the purpose in hand. In this way, our work proceeds from the original idea or some standpoint based on certain values.

This form of division into levels is described in so-called *theory of logical types*[69]. Here there is strict requirement to avoid placing an *element* at the wrong level. It is illogical and makes communication and problem solving difficult. By placing the appropriate things at the appropriate level, and knowing, moreover, what level we ought to be at, on different occasions, we have more favourable conditions for finding a solution. When level A lies above level B, A is said to be a *metalevel* in relation to level B.

If we pile a number of abstraction levels on top of each other, we get a diagram which ends in concrete action or activity (Fig 5.1).

Chapter 5

When we speak about levels of abstraction, the most abstract level is to be found at the top.

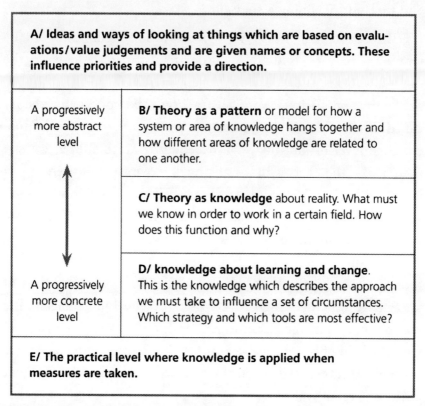

FIG. 5.1 *Levels of abstraction according to logical type theory*

Ideas and ways of looking at things

If we describe our area of activity in terms of theory and practice, we risk missing the abstraction level concerning ideas and ways of looking at things. This uppermost level is more philosophical in its content and perhaps as a result more difficult to bring out in a discussion. Nevertheless this level involves both useful and important dimensions of an activity. At this level, ideas arise and are conceptualised. One example is Antonovsky's idea of salutogenesis which we have described earlier. This idea, through its appeal has led many people to discuss what health promotion is. In this way, the idea has grown into a field of interest. It is made up of many different professions and disciplines each of which contribute their knowledge and experience. As a result, we can also describe the idea of salutogenesis as a field of knowledge under development, the salutogenic health promotion's field of knowledge.

Ideas and the reflections they give rise to, can also be used as a starting point when we have to plan and carry out some activity. Different ideas are compared and evaluated on the basis of the value or effects that their application can create. On the basis of the idea, a vision can be formulated as a normative guide. The next step can be to describe a purpose, the answer to the question "why". It is the good ideas which can become attractive visions and goals of human endeavour. When the visions and goals are based on values which appeal to people, this leads to greater involvement and loyalty with regard to the task in hand. Research has shown that good ideas and sets of values have great importance in ensuring that workplaces and organisations are healthy and sustainable. The work of Collins and Porras[70] in comparing the core values of American companies and investigating how far the respective organisations have exhibited sustainability, is particularly well known. Corporate managements which expressed good ideas and values which appealed to people in the organisation, had the best economic performance in the long term.

Human values become of political interest, when different interest groups come into conflict with one another. In an organisation, there is an open stage for political influence which fills up with people when visions and goals are to be formulated. Often there is a hidden agenda for the political game, which according to research into organisations, is both more powerful and difficult to control. According to Bolman & Deal[71], the *political perspective* is an important explanation for how organisations function. This perspective has great importance for the values which influence the positions and attitudes people adopt and their behaviour. It is possible to lead with the help of values. At the same time, everyone ought to reflect about the leading values in e.g. a workplace.

Theory as pattern

When people carrying out research investigate a part of reality, they sometimes draw general conclusions from what they have observed. For example, people who study communication in a work group observe certain behaviour and patterns which can be assumed to be typical for people in a group. These behavioural patterns, which are found in the specific group studied, are presumably also to be found in many other groups or indeed in all groups. In a report that others can read, the researchers describe their theory, which is true as far as the group studied is concerned, but a hypothesis with regard to all other groups. It then becomes possible for other people to test to see if the theory holds for other groups. One way of doing this, is to try when critically testing the theory, to see if it is refutable. If it cannot be refuted, the probability that it is correct increases. If, as more and more groups are tested, the theory is found to be confirmed, the more certain the theory becomes.

When we have a theory which has survived successfully a sufficient mumber of critical tests, it becomes a paradigm or model which helps us to generate new knowledge, whence the saying that "there is nothing so practical as a good theory". Certain theories aim at giving more overarching explanations of a phenomenon and can at their *metatheoretic* level serve as a guide for other theories. One example of this is Antonovsky's SOC-theory which is recommended in the present book. The theory of a sense of coherence is a concept or pattern which sets out at a more general level to understand what is important if human beings are to preserve or even improve their health. This theory was initially a hypothesis based on a study of the life and health of some forty subjects. After tests[72] by many other researchers, the theory has been shown to have great generality.

Theory as knowledge

It is at this level of abstraction that we find the knowledge which describes or explains the situation in which we find ourselves and the things or systems we are working with. This understanding is a prerequisite for being able to formulate problems, analyse, judge, discuss and reflect, evaluate and compare what we are working with.

The significance of learning and the role of knowledge in working life and social development, have for several decades been central issues. Lifelong learning and the learning organisation are merely some examples of concepts which point to the importance of knowledge today. This is true not only in the case of the development of products and services. Learning has also become an important way of bringing about change in organisations and systems of work. The more members of a workforce who possess knowledge and want to use it, the more clever brains there are to raise questions and come with proposals about improvements, even when it concerns organisation and other working conditions. Some concepts linked to knowledge and learning may be worth mentioning in this context. Knowledge is to be found in different places. In part there is scientifically derived knowledge within various disciplines which is presented in educational contexts or which is available in books and articles. There is also everyday knowledge which is being developed all the time by every active individual who has an interest in what he or she is doing. A good way of developing competence is to combine both these sources of knowledge by reflecting in everyday life about our practice with the support of theory and experiences which others have written about. Ellström[73] holds that this reflexive

Chapter 5

learning provides a deeper understanding and increases our capacity for knowledge-based action.

Another aspect concerns the form of knowledge we need. Are specialists or generalists best fitted to deal with a certain task? Should we devote ourselves to learning a great deal about a small area of a subject or is it more important to know a little about a larger area? How specialised must one be to understand, interpret and take decisions about various issues? In terms of system thinking, we can describe the whole of existence from the human cell at the lowest level to the whole universe as the great overarching system. Perhaps it is logical that the person who is responsible for the working environment and health questions at a workplace, is a generalist with general knowledge about people and how organisations function. In addition, we also need people with specialist knowledge within particular fields such as human physiology, psychology, working environment, management etc.

It can seem to bed for conflict when the Western research tradition celebrates a specialisation whereby people seek to deepen their expertise and are forced to set limits to their field of knowledge, while at the same time, the capacity for an overall view, a feeling for the matter as a whole and co-ordination are so important in a developed society. It is crystal clear that there is a need both for generalists and specialists. The collective results in problem solving and development work can be promoted through close contacts, mutual respect and a capacity for co-operation.

Aaron Antonovsky has made several contributions to the development of health promotion. He put forward the idea of salutogenesis and he also argued for greater scope for the social sciences in health promotion. In an article which appeared 1994, Antonovsky[74] maintained that the sociology of health was a discipline which forms an integral part of a salutogenic approach to health promotion. He held that

> "The voluminous writing of the holistic approach to health shows a near-total absence of the larger social system in which the mind-body relationship operates".

Knowledge about learning and change

It is not enough simply to have a large store of knowledge: we also have to be able to make use of it. Ellström[75] defines competence as "an individual's capacity for dealing with a certain task, situation or work" (p. 149). A capacity for dealing with things implies the ability to solve problems and bring about change. This implies that the individual can use and combine knowledge, intellectual and manual skills, social skills, attitudes and personal qualities which are concerned with cognition and creativity. Ellström stresses the importance of linking theoretical knowledge with practical experience. Purely theoretical training has been shown to be difficult to transfer and apply, whereas experience-based learning tends to be dependent on the concrete situation and is difficult to make use of in new or unknown circumstances.

In describing two strategies for change, Flemming Norrgren[76] holds that learning and competence to deal with change are important prerequisites in working life. The first he calls the *programmatic* strategy which is characterised by careful planning, steering and control. Most things are thought out and decided in advance, preferably from a *top-down* perspective.

The second strategy Norrgren calls the *learning strategy*. This entails that maximum attention is focussed on the run-up to, as well as the actual course of the process of change. Workforce participation, dialogue and equality are the beginning to an analysis of the problem and to suggestions about how it is to be solved.

Chapter 5

A member of the workforce is an actor and acquires a growing understanding of where the company is headed: the learning process has begun.

It is contended that the new forms of organisation we see today, presuppose continual lifelong learning. "lean production", "just in time" and "continuous improvement" are examples of phenomena, primarily in industrial production but also in the service sector, which rely upon decentralisation and individual competence to solve problems and bring about change.

Ellström points out the paradox in modern organisational applications. His view is that these customer-led organisations tend towards reductions in staff with the result that the work involves greater stress, which in turn reduces work motivation and makes learning more difficult. It is important for the management of the organisation to combine a high technical level of development with a working organisation which promotes learning and develops the competence of its workforce.

In his book, *The Corrosion of Character*, Sennet [77] describes how although the modern organisations certainly give greater influence, personal responsibility and more learning to the worker, freedom and individual responsibility can easily become a straightjacket: an employee is free to work as much as possible. Being responsible for checking one's own work, high motivation and scope for personal initiative are positive, but can also lead to frustration and an increase in stress. The competent and autonomous employee runs the risk of being exposed to "burn out" and other destructive biological mechanisms.

Learning which leads to new knowledge gives the individual increased understanding, a capacity for problem solving and participation in decision-making. The French philosopher Foucault points out that a positive form of learning leads to *an increased power over one's situation.* He holds that knowledge can give power or at least the possibility of becoming acquainted with power. Foucault also maintains that learning can also be a way of exercising control. People can be "taught" or "brought up" to fit in to a certain culture or a certain "sheepfold" which somebody else controls. What is important in the learning process then becomes, not the individual's situation and possibilities, but rather those of the system, organisation or society. It becomes an ethical question how far individuals should have a learning opportunity which does not aim at subordination, but instead at increasing their possibilities of participating and extending their control over their work situation.

Bearing in mind this critical reflection, we can state that learning and the competence to deal with change, have great importance in contemporary society. It is also indispensable in the case of health promotion, viewed as a strategy for change, in bringing about improved conditions for health in the workplace.

We must also remember that problem solving capacity and competence to deal with change, often are dependent on a form of process knowledge where the individual has the capacity to produce the right knowledge and to use the right tools in the right order in order to find the best solution. This is a knowledge of how the practical side of things should be organised and carried out in order that an activity functions properly. Knowledge of methods, their content, application and effects, is needed in order to be able to choose the right method in the right circumstances and to design programs and processes in a systematic and effective way.

Today, learning is a prerequisite for change. Change is a process which requires knowledge of how processes should be designed and steered. In the process, "tools" and methods are employed which are appropriate to the end in view. Knowledge about all this

Chapter 5

is both general in character and specific to the type of work and the circumstances we have to deal with. In a field such as health promotion, this form of work has an important place, since there is a central focus on bringing about change.

Summary

There is much more to say about knowledge and views of knowledge, and about change and learning than we have managed to say here. This chapter has set out to describe the importance both of having a view of knowledge and a knowledge base when it comes to working within the area of health promotion. If we are to argue for and then use a certain strategy for health work, we can afford to devote some thought to theory in considering the following four points:

- What is the idea that forms our starting point and what values follow from it? (E.g. the idea of salutogenesis on the one hand and pathogenesis on the other).

- Is there an overarching theory or a explanatory pattern which can serve as our guide when we decide what knowledge is needed in order to understand how a dimension such as human health or a complex system such as a workplace functions? (E.g. SOC theory, systems theory, theory of logical types.)

- What knowledge is needed about the different parts of the system? (E.g. psychology, physiology, organisation theory).

- What branch of knowledge explains how the practical activities should be dealt with and how change is brought about? (E.g. pedagogy, didactics, the theory of change.)

From idea to theory

It is only after this series of issues has been dealt with, that the knowledge of methods and the way of doing things enters the concrete work. It is sometimes far too easy to pass directly to action, when one sees a problem or has been given a task to carry out. Presumably the work of health promotion will lead to better results if idea, theory and practice are combined.

A theory in the sense of a pattern or model to describe and explain something's construction or how it works, is of great assistance, both in further research and in shaping practice. Since different scientific disciplines have different theories, it is the task of health promotion to discover its own guiding theory or theories. In the current literature, a proposal has been made for a guiding theory for health promotion on the basis of two questions:

1. *What conditions have to be satisfied in order for health to be preserved and improved?*

2. *How can we bring about a change which will lead to better preconditions for health?*

Antonovsky has put forward a theory which serves as a guide with regard to the first question. In the next chapter we shall describe this Sense of Coherence (SOC) theory[78]. How has it arisen? What does it contain? Finally what significance does it have when we are faced with transforming health promotion from a good idea into an effective practice?

6. The mystery of health and sense of coherence

In this chapter, an overarching theory is presented which can help us to describe the preconditions which are important in order for health to be preserved and improved. We are now speaking of theory as a pattern about how different constituent parts of reality or some area of knowledge are interlinked. What connections are there? What different groups of factors promoting health can be described? Are there obvious common characteristic features for a group of health factors which can assist us when we have to identify several which have the same properties? A theory about the connections which influence human health can be useful when, for example, we investigate if a workplace has access to one of several factors which satisfy a certain need or type of criterion for promoting health.

When new knowledge arises in a certain area, it is frequently based on earlier knowledge which researchers arrived in the past and wrote up in scientific journals and dissertations. At every point in history, there is a certain knowledge and body of theory which is accepted by the scientific establishment. Textbooks in the respective fields are written on the basis of current knowledge. Some truths have a long life and are important for several generations of researchers. Others are more short-lived and collapse when new knowledge shows that truth is to be found elsewhere.

When new knowledge in the form of theories or explanations pops up somewhere, it must be accepted by the scientific associations in the respective discipline. Naturally a competitive situation arises when new advances in a discipline occur, raising questions about -

Chapter 6

or attracting attention away- from the established truths which have hitherto been accepted. Some of the scientific ideas and applications which have appeared in history have been more radical and led to extensive new thinking within a discipline In the history and philosophy of science, these new, more revolutionary, traditions of thought are called "paradigms". Thomas Kuhn[79], who coined the term, describes a new paradigm as a new type of thinking which defines for several generations of scientists within a given area of research the problems and methods which are legitimate. The new paradigms also share two characteristics:

> Their achievement was sufficiently unprecedented to attract an enduring group of adherents away from competing modes of scientific activity. Simultaneously it was sufficiently open-ended to leave all sorts of problems for the redefined group of practitioners to resolve. (p.10)

The question arises whether Antonovsky's idea of salutogenesis means that future researchers will be able to state that a new paradigm in health research popped up in the 1970s. Does the idea of salutogenesis entail a radically new way of framing questions and in research, of seeking explanations about the origin and causes of health? Whether it turns out to be a paradigm shift or not, there is much nevertheless to suggest that interest in Antonovsky's idea and theorising is considerable. Thousands of articles about salutogenesis and the theory of *Sense of Coherence* (SOC) have been published since his first book appeared in 1979[80]. The idea and theory has been tested in several fields such as social work, stress research and work dealing with change. Many have also tested both the idea and theory in working life and have found that it is valid both as a strategy and as an explanatory model for how we can promote health in the workplace.[81]

The great interest in both testing and critically examining Antonovsky's theory is a strength when we want to depend on it as a guide in health promotion. It has been carefully tested that sense of coherence is really a salutogenic theory of health. We know that it is plausible and reliable in serving as a theory for describing what leads to health.[82] Critical articles about how Antonovsky's theory has been used, also help us in finding a relevant application of it within the framework of health promotion in working life.[83]

Who was Aaron Antonovsky?

Aaron Antonovsky was born in the USA in 1923, gained a doctorate in sociology at Yale University and emigrated to Israel in 1960. At the end of the 1960s, he became known as a leading medical sociologist. In 1967, he published an article[84] which was to exert great influence in stimulating interest and insight into how health varies between different social groups in a society. In this way, he contributed to a realisation that human health is not simply about body and soul and the immediate living environment. Both the structure of society of which we form part and the extent to which we succeed in living our life in society, have a crucial role for the individual's well-being and for the average rate of illness and mortality from a public health perspective.

From 1972, Aaron Antonovsky worked at Ben-Gurion University of the Negev. He taught physicians and students in the medical faculty the importance of looking at human beings in their particular situation and of bearing in mind that health consists of several dimensions. Simultaneously with his teaching, he developed his ideas and research on the basis of the connection between stress and ill-health, also making use later of the salutogenic model and the health continuum.

Chapter 6

The origin of the salutogenic idea in Antonovsky's research was a study in 1971, which focussed on 1150 Israeli women and dealt with their adjustment to the menopause. He happened to ask if they had been in concentration camps during the Second World War and of the 287 women of European origin, 77 replied "yes" to the question. In connection with the purpose of the study, he discovered, as expected, that fewer of the former concentration camp prisoners had adapted well. But having stated this, it struck Antonovsky that this self-evident result concealed the knowledge that despite the difference between the groups, there were remarkably many of the concentration camp inmates who had coped well. This gave rise to the salutogenic question: how is it that certain people, despite terrible experiences such an internment, war, flight to another country and other stress, nevertheless enjoyed pretty good health and a happy life? There was one question he was keen to answer: what are the resistance resources which allow people to manage this? Seeking an answer to this question was to become Antonovsky's central task during the next 23 years.

After the publication of his first book *Health Stress and Coping* in 1979[85], there was accelerating interest in his work. Among other things, he was awarded an honorary doctorate by the Nordic School of Public Health in Gothenburg in 1993. After his retirement in 1992, he continued to give lectures and carry out research until his death on 7 July 1994.. According to Antonowsky's own testimony, his wife Helen who was also active at Ben-Gurion University, was of great importance for his work.

Antonovsky's salutogenic model as a theoretical guide for health promotion

The idea and inspiration for this section heading comes from Aaron Antonovsky himself. In an article which appeared

The mystery of health and sense of coherence

posthumously in 1996[86], he described how development in health promotion ran the risk of stagnating because of the lack of a theoretical guide.

> "The concept of health promotion, revolutionary in the best sense when first introduced, is in danger of stagnation. This is the case because thinking and research have not been exploited to formulate a theory to guide the field." (p. 11)

He held that the illness perspective and thinking in terms of risk factors were far too dominant. Even proponents of health promotion therefore end up with the dichotomous –either/or- classification of people into two groups -one which consists of people who are negatively affected by illness and the other consisting of people who manage to cope. If instead we were to adopt a continuum model which is based on the idea that every human being, at every moment in time, finds himself or herself located at some point of the continuum between the two poles of health and ill-health, we would obtain an approach to health which mirrors reality in a much more plausible way. By asking the salutogenic question- how is it that people move towards the health pole of the continuum, we find that all people can be an object for health promotion activities. Irrespective of where we find ourselves on the continuum and of who we are, whether ill, handicapped or very healthy, there is an opportunity to reflect on how a movement towards the health pole can take place.

It is also possible to link a robust and lucid theory of health promotion with this concept. In his article, Antonovsky held that research on the basis of a salutogenic approach needed to be strengthened and practical application required a concept to guide it and in this matter the theory of a sense of coherence (SOC) was a model which could be further developed.

Chapter 6

There are many excellent suggestions about the factors which are positive for health. There is an increasing number of researchers who are interested in identifying the determinants of health in different contexts. During the beginning of the twenty-first century, great attention has been paid to the health factors in working life. However, it would seem that this increasing listing of the determining factors of health is insufficient to bring about better health. In his 1996 article, Antonovsky was of the opinion that there was a need for a theory to bind together all these excellent proposals, in other words a metatheory which can form a pattern and a guide, both for practitioners and research workers.

The collection of health factors which started Antonovsky off on his quest for a theory which could explain the movement towards the continuum's health pole, was made up of what he called *general resistance resources* (GRR), In his research[87] into social class, poverty, and health, he had discovered factors which were significant for health. These were factors which collectively formed a general resource in people, enabling them to withstand, cope and survive all the stress and dangerous pathogens which life is full of. These general resistance resources are to be found in people's historical background and growth and in their social and cultural surroundings. According to Antonovsky[88], apart from hereditary qualities, they consist of things which are tied to material welfare, knowledge, self awareness, coping capacity, social networks, commitment, cultural stability, fantasy, religion and health awareness.

Antonovsky was not content with merely having a long list of resistance resources or factors which are good for health. He wanted to have a theory which would explain how these factors were linked together. What unites them? How do they exert an effect and how do they co-operate and interact?

The mystery of health and sense of coherence

What seemed to be common to all these factors was that they helped to give experiences of life which quite simply helped people to make sense of their existence.

> What united them, it seemed to me, was that they all fostered repeated life experiences which, to put it at its simplest, helped one to see the world as "making sense", cognitively, instrumentally and emotionally" 89. (p. 15).

This collection of resistance resources along with positive life experiences and the successful management of stress formed the basis of what Antonovsky called Sense of Coherence (SOC). In his first book, he discusses at length how the GRR combine to create this sense of coherence in the individual. They can also help to bring about a movement across the continuum toward the health pole.

In order to arrive at a deeper insight into what characterises SOC, Antonovsky carried out a study[90] in which he interviewed 51 people who all shared something in common, namely they had experienced a severe trauma in their lives, but were considered to have managed to pull themselves through it successfully. When these people described how they looked upon their lives, Antonovsky discovered three central themes which subsequently formed the foundations of his theory. In his analysis of the interviews, he found repeatedly that individuals who had coped successfully with stresses and crises had an experience of *comprehensibility, manageability* and *meaningfulness*.

In his book *Unravelling the mystery of health* [91], Antonovsky reviews this part of his work where after describing resistance resources, he further develops the idea of salutogenesis and the continuum approach in order to formulate a clear and readily understandable theory about the prerequisites which appear to be crucial for the development and maintenance of health. He identifies and describes the three concepts, how they relate to one another, and their limitations.

Perhaps it is just this triad, the "holy trinity", which has made SOC so interesting and attractive as a theory. It is easy to remember and easy to translate into operational terms in various everyday situations and expresses simultaneously a life's wisdom which can help people to cope with their situation. Antonovsky himself summarises SOC with his three subsidiary concepts of comprehensibility, manageability and meaningfulness in the following way:

> ...a global orientation that expresses the extent to which one has a pervasive, enduring though dynamic feeling of confidence that (1) the stimuli deriving from one's internal and external environments in the course of living are structured, predictable and explicable. (2) the resources are available to one to meet the demands posed by these stimuli; and (3) these demands are challenges, worthy of investment and engagement. (1987 p.19)

Comprehensibility

According to Antonovsky, comprehensibility is bound up with the extent to which one can perceive all the information and stimuli in one's existence as something graspable, structured and predictable. It entails that the individual has a stable capacity to judge reality and understand why it is, as it is or becomes what it becomes. This control dimension is a fundamental prerequisite for the next concept, namely manageability. We must know how matters stand and how problems can be attacked in order to cope with a problematic situation. People who experience the world as disordered or chaotic, have probably difficulty in seeing how they can deal with their situation.

Manageability

Manageability is based on the idea that we have the feeling that there are resources at our disposal and that it is possible for us to act on the basis of the demands that are placed on us. It can be a question of our own resources which we personally control or those of people near to us.

The concept of manageability (or similar expressions) to describe human autonomy and the possibility of steering our own lives, are common in various stress and coping theories. One of the best known and cited is the Karasek- Theorell[92] demand/control model. This theory is considered fundamental for understanding how a situation leads to tension and stress.

Kobasa[93], another stress researcher, holds that certain people are tougher and more robust in stressful situations where (among other things) the degree of control, in the form of influence over one's situation, is an important factor.

Manageability in the SOC theory differs from other traditional models of coping in the sense that Antonovsky's concept of manageability also covers human beings' propensity to be able and willing themselves to use their surroundings to help them. It is an asset to recognise resources in our fellow-humans and to be humble enough to ask for help and support. To rely exclusively on ourselves gives a more restricted definition of manageability or the ability to act. The concept of manageability in SOC also includes the belief that we can cope with the demands or difficult situations which we encounter in life, by relying on ourselves *and* on help from others,. Good manageability means that instead of feeling ourselves as passive victims of catastrophes, big and small, it is possible by taking a grip of things to ride through life's storms.

Chapter 6

Meaningfulness

Meaningfulness gives the motivational ingredient of SOC theory and answers the why-questions of life. Why should I do this? Why does this happen? What does this give me? What does this give other people in the world as a whole? When we are faced with a task, there are two ways to react: either the task is considered as a heavy burden or we look upon it as an inviting challenge which is worth becoming involved in.

The experience of meaningfulness has a firm emotional basis in human beings. Sometimes, commitment, energy and joy are elicited without us being able to directly say why. How is it that a workplace consists of people who experience just that, while another comparable workplace consists of people who consider their duties tiresome and meaningless, even hopeless. ?

One person who has analysed the importance of meaning for human beings, and who has also inspired Antonovsky is Viktor Frankl. He provides many examples[94] of what creates meaning and what the meaningfulness issue means for the lives and survival of human beings. Frankl addresses the importance of relationships, the actions we take, a goal to strive towards, aesthetic experiences and so on. He himself was in a concentration camp and asserted afterwards that those prisoners who could find a meaning in what awaited them, were best equipped to survive. Frankl employs Nietzsche's words to describe this: "a person who has an answer to "Why live?", can withstand almost every How"[95].

Frankl describes the importance of the issue of meaningfulness both in the everyday sense where the day's minor events, experiences and relations create meaning, and in the sense of having a more far-reaching and overarching feeling of meaningfulness. He gives examples such as religious belief, a life task or a longing after something which lies far away.

Of the three constituent concepts which make up SOC, Antonovsky holds that meaningfulness appears to be the most important. Without it, comprehensibility and manageability are likely to be of rather short duration. In addition, it is probably also the case that people with a strong feeling of meaningfulness and motivation, will acquire both knowledge and resources to solve their tasks.

The parts and the whole

The three concepts make Antonovsky's theory a model which, in its entirety, finds scope for a successful ability to deal with problems which arise in the human encounter with the stresses and strains of life. Antonovsky holds that a strong SOC is a decisive factor in influencing the movement along the continuum towards the health pole.

This theory allows a further perspective with regard to the health issue, since the factors which are important in promoting human beings' health are not to be found simply within people themselves. The social circumstances at different levels of the system, the surrounding environment and the temporal perspective -the past the present and the future - are important for the *Sense of Coherence*.

Antonovsky describes SOC as a global orientation and as a "dispositional orientation"[96]. By this, he means that SOC is a capacity which the individual has in addition to personality characteristics and coping strategies. SOC should be seen as a fundamental resource in the individual which does not vary from one setting to another. It is the individual's SOC which is the basis for the choice of coping strategy. A strong SOC gives an increased capacity to choose and operate with appropriate coping strategies in a flexible way. In the case of a weak SOC, the individual tends to let personality

Chapter 6

characteristics and temporary feelings steer the choice of coping strategy, irrespective of the demands which the situation makes.

> In fact it is precisely the person with a weak SOC- confused, unsure of resources, and wishing to run away- who is likely to allow personality traits and tendencies to determine behaviour, irrespective of the nature of the situation"'. (p. 57)

A strong SOC increases the chance of coping successfully, which in turn gives rise to an experience which further reinforces the individual's SOC.

The properties of SOC prompt many other reflections. If one tries to order the constituent properties of SOC in terms of importance, does meaningfulness come first? Or is meaningfulness a first fundamental prerequisite for the other two? Is my sense of coherence something which varies between different sectors in life? Is the feeling of coherence a basic strength in life which is formed during my growth to adulthood, so that, as an adult, I shall be steadfast and not easily pushed off course? Or is it the case that the sense of coherence can be related to a certain situation or certain circumstances e.g. my work and there it can be influenced and improved if management, the working environment and other conditions are good? These questions serve as a reminder that there is still a lot of research to be done, when we want to know what it is that influences the sense of coherence and how it is accomplished.

Testing a theory's validity

Inductive research consists in studying empirical reality and on that basis, formulating a theory. Antonovsky's theory of a Sense of Coherence (SOC) is based on empirical evidence derived from responses to interview questions. Antonovsky also carried out *deductive* research in the sense that starting from a theory, he

The mystery of health and sense of coherence

devised a questionnaire and proceeded to test his theory on large groups.

Note that we travel in two different directions in developing our theory and in using the theory. First, we study empirical reality and formulate a theory on the basis of our study. On the basis of the information acquired, the researcher then makes a generalisation and claims "This is how it is.. here's how we can explain in general terms the phenomenon I have studied. This is my theory.".

Researchers also take the *deductive* route from theory to reality : they have to test to what extent their theory is valid. Thus the theory is a hypothesis, that is to say, an assumption about how things are and their general validity. Researchers formulate their questions on the basis of theory and test if the answer is the same as in the first situation, when the theory was first formulated.

One can say that as practitioners we also use the latter deductive approach when we have a theory as a pattern for interpreting reality and for planning our activity, using some theory to guide us in what we do and how we do it.

Antonovsky's Sense of Coherence as a theory of health involves all these aspects. First of all it is derived from empirical reality, secondly it is tested on many groups in different cultures and thirdly it has been shown to be useful in a multitude of ways in guiding practice.

A newly formulated theory constitutes a hypothesis that something or other can be explained in a particular way and possesses general validity. The hypothesis must therefore be tested with respect to different groups and situations. On the basis of the Sense of Coherence theory, Antonovsky constructed a questionnaire containing 29 questions. He wished to study to what extent it is true that people, who have a high sense of coherence, also have good health.

Chapter 6

Antonovsky personally carried out the testing of the theory's validity and this was repeated by many after him[98]. The results are nearly unanimous in concluding that SOC with its three subsidiary concepts is a valid theory or model for health. It is fit for use and we can rely on it to guide us when we are looking for the determinants of health e.g. in working life.

Naturally there is also criticism of Antonovsky's theory and how it is used. Geyer[99] holds that the SOC theory does not contribute anything that is new: it measures the same phenomenon as the questionnaire for depression and anxiety, but with an inverse correlation.

Tishelman[100] considers Antonovsky's original questionnaire to be more of a psychological instrument than something which can describe the global orientation on the basis of social and cultural aspects, as Antonovsky maintains that SOC does. According to Tishelman, this also explains why Antonovsky's questionnaire is often used wrongly as a psychological tool, in spite of the fact that the theory is sociological in character, with greater emphasis on structural factors and coping resources at a meta-level, than on psychological features of personality.

The critical examination is important and as far as SOC is concerned provides no reason for rejecting the theory. It serves very well as a model in the search for health factors in a particular setting, for example in the workplace. We can carry this work out at a general level in order to find out what is important for health in working life in general, and to determine the general aspects of comprehensibility, manageability and meaningfulness which working life can supply. Moreover we have the unique situation of a particular workplace.

The use of the whole of the social and technological context as the explanatory basis of health in the workplace leads to very great complexity. The situation in each workplace, at each moment of time, involves its own unique collection of explanations of what it is that influences health. It is therefore necessary to pose the following problem also at the local level: how do we meet the demands for comprehensibility, manageability and meaningfulness in our particular workplace?

Such a way of formulating things provides us also with a question which can serve as a starting point in interpretively and pedagogically directing an activity. How are we in the management to organise and create the prerequisites for a high degree of SOC in our workplaces? Supervisors and top management can ask themselves the following question: how should I manage things to provide high degrees of comprehensibility, manageability and meaningfulness for my workforce?

Antonovsky encouraged other researchers and practitioners to continue with the testing and development of his theory of the Sense of Coherence. He also proposed in the article cited above "The salutogenic model as a theory to guide health promotion" the use of his theory as a way of organising the work of health promotion.

It is this which forms our starting point when we use both Antonovsky's idea of salutogenesis and his SOC theory as a model to guide us when we move on to develop health promotion as a strategy for health. We have presented the idea of salutogenesis as a self-evident perspective for health promotion thinking and it is essential if we are to achieve success with health promotion as a practical method of bringing about change and development. Antonovsky's Sense of Coherence plays the part of an overarching model, a metatheory, which can help us to design "healthy" organisations which bring about conditions which benefit both people's health and the performance of the enterprise.

Chapter 6

The Sense of Coherence theory, however, cannot solely by itself provide a patent solution on how to carry out health promotion. As we have remarked previously, SOC can guide us in describing how things should be in order to ensure that health is preserved and improved. It is the answer to the question: *what preconditions must be satisfied for health to be preserved and improved?* We must also complement this with a theoretical model which helps to answer the question: *how do we achieve a change that will bring about better preconditions for health?* We shall return to this latter question further on in the book.

From "global orientation" to everyday level

A theory is a picture or an abstract explanation of reality. What we are going to discuss here is rooted in the popular saying that "there is nothing so practical as a good theory". In other words, the extent to which a theory is good or bad, is determined largely by how far it can subsequently be translated and applied to empirical reality. We shall take up certain aspects of the application of SOC and discuss how we can argue and reason, using them as a basic frame of reference.

Antonovsky describes SOC as a "global orientation", holding that a Sense of Coherence is a universal human characteristic, an attitude and way of reacting to things which is largely formed during our growth to adulthood. It is useful at all the stages of life and is the basis for our capacity to discover appropriate strategies in dealing with the problems we encounter.

By the time maturity has been reached, the individual's level of SOC cannot be influenced very much, especially in the case of individuals with a high level of SOC. The basis which we have laid down during our growth to maturity remains with us and, according to Antonovsky, we have no alternative later on, other than to be content with what we have. He also points out that those individuals who have a weaker SOC, have greater possibilities of improving it. If individuals are part of contexts which create a sense of meaningfulness, if they participate in experiences which increase their sense of comprehensibility and if they make courageous decisions which reinforce their view that difficult situations are challenges which they can manage, all of this tends to increase their SOC.

SOC is a human quality which helps to preserve health or even sees to it that it is improved. The goal of health promotion work which employs SOC as a model, ought to be to create situations which help to ensure that human beings' Sense of Coherence is preserved and in the case of those with a weak SOC, that it is strengthened.

The first step in making SOC useful, is to discover concepts which show what comprehensibility, manageability and meaningfulness consist of, in everyday situations, e.g. at work.

Everyday comprehensibility

If we are to discover what comprehensibility means in practice, we need to have some knowledge about the world in a wider context, where we see our own place in relation to things as a whole. It is a question of knowing what there is and what is happening and why some event evolves as it does. Verbal capacity and verbal levels are brought into play in order for us to understand what all the information that reaches us, means. How easy is it for an immigrant to acquire comprehensibility when communications are made up of words in a foreign language and is accompanied by bodily language, facial expressions and an intonation which gives the spoken speech its real meaning? Do I understand what my workmate is saying?

Chapter 6

What does the person mean? Do they wish me well or is my position being questioned?

Even my history or our history is a source of comprehensibility: not primarily the kings or wars that have affected the world, but rather the more intimately related history of my own country, my own district, my own family and my own workplace, which can help to increase comprehensibility about why things are as they are just now. A historical foundation can also be seen as a form of roots which create a relationship between the present world and the past. Such a foundation is an important contribution to the Sense of Coherence in life.

Comprehensibility at the workplace level is to see the connection between different parts of the organisational system. What does the overall work process look like in the company? Where do I fit into it and what function have I in the large organisation? What does our organisation look like, not merely on paper, but as it is in terms of living people on physical premises? This kind of overview is easier to acquire in a small organisation.

Comprehensibility is also very much about the capacity to assimilate information, to interpret what we hear or see, and to draw conclusions about cause and effect. The individual's intellectual capacity in terms of memory, logic, and creativity are significant in determining the level of comprehensibility that he or she can achieve.

What effects arise from low comprehensibility? Not to know or understand leads to uncertainty and alienation. We are less likely to make wise decisions and do the right thing, when we lack knowledge. The ability to defend and safeguard our own needs and rights decreases for those who do not know how society is organised or how organisations are structured.

Everyday manageability

Manageability in practice deals with the capability and opportunity of individuals to influence their situation and their surroundings. The concept of *empowerment* is often employed in Health Promotion literature[101] as a convenient expression for everyday power and access to the resources people need, in order to be able to manage their lives in a way which promotes wellness. There is a strong ideological component in this concept which emphasises the right of individuals to make decisions about, and take charge of their own lives. Individuals are active participants, instead of passive recipients of the decisions and conditions relayed by society.

In order to achieve manageability, comprehensibility is needed since we have to know what has to be done and how it should be done, before we can act. Knowledge of all kinds is needed, and the more complex the situation, the more knowledge is needed to be able to cope with managing the situation and the tasks in hand. Pedagogy and learning have therefore an important role in the health promotion approach to work.

In addition to understanding the situation, manageability is also based on different kinds of resources and empowering circumstances. Manageability in the workplace is based on possessing professional skills for the job and professional skills which arise from experience. Personal resources such as communicative ability, the physical capacity to cope and physiological motor skills can also be valuable for manageability.

In the workplace, manageability is created in the form of good tools and the right material. Helpful workmates and a well-functioning organisation also make a contribution to manageability. The ability to manage our situation at work, is also based on the fact that as individuals, we are allowed to deal with things to the extent

Chapter 6

that we are able. Influence over our work conditions, the possibility of regulating the pace of the work and the opportunity to exercise our own initiative, contributes to autonomy and manageability.

Privately, manageability can be a matter of access to money, people or things which I need to achieve goals having priority. Problem solving capacity, personal energy and staying power, as well as imagination, can be different forms of resources which are needed to cope with the tasks that I am faced with.

Manageability can also be a matter of the individual's capacity to evaluate and re-assess the importance of a situation. It is possible, for example, to succeed in dealing with a task which appears to be beyond one's powers, by reassessing its importance. How important is it really? What can reasonably be demanded from me, given my capacity and my resources? What is the worst that can happen to me if I cannot succeed in the way that I wish? People who are able to stop thinking about a difficult problem, which they cannot, in any case, solve with the resources at their disposal or given their current situation, also display good manageability.

What are the consequences of low manageability? When the tasks and the demands become too many or too difficult, manageability decreases and people become frustrated. Here, we can apply the stress researcher's explanations of why the degree of autonomy or scope for action in relation to demands or stress, is so crucial in determining whether a person is afflicted with negative stress. There must be some kind of balance between the demands made and the resources which individual disposes over.

Everyday meaningfulness

Objective is the term we assign to the picture of some state in the future which gives human beings the motivation to act and to begin doing things which will lead to what is motivating them.

The mystery of health and sense of coherence

People, who are motivated, have the energy or drive to act and do something. People who lack motivation have neither the physical or mental capability and energy nor the desire to make a contribution to achieving the result. Antonovsky describes meaningfulness as the motivational component of SOC and asserts that because of that role, it is the most important of the three concepts of comprehensibility, manageability and meaningfulness. If the question of meaning is left unanswered, interest in increasing one's comprehensibility and one's motivation to deal with the problem, decreases. The experience of meaningfulness and *joie de vivre* are intimately linked.

We perceive a person totally devoid of the experience of meaning in his or her life as listless and unenterprising, quite uninterested in the invitations and challenges of their surroundings. A day without meaning feels like a lost day. A life without meaning is empty and hopeless; both body and soul are devoid of will and soon health is affected, unless some experience that gives meaning pops up. We can also see people with a strong sense of meaningfulness, whose eyes shine from a deep sense of involvement and presence, in order to absorb all the impressions that the world about supplies. There is both determination and energy to do something, since motivation becomes stronger the more meaningful the situation is experienced.

Quite concretely, we can say that meaningfulness is a positive motivational component which can, to some extent, arise when the motive is presented in the form of an attractive vision or goal. It can also arise from small events or situations which satisfy some important temporary need. The experience of meaningfulness is relative, depending upon my emotional state and how much is needed to go from an experience of emptiness to one of meaningfulness.

When we find ourselves at a "low", it needs only a few small words or actions to transform the day into one that is meaningful;

Chapter 6

a fellow human being can, by their mere presence, contribute to someone's feeling of meaningfulness.

What is needed to create the experience of meaningfulness must therefore be a question which arises in relation to individuals' needs and their values with regard to the world around them. If people have their personal, fundamental needs satisfied, there is a greater chance that they will be motivated to contribute to the collective goal. Collins & Porras[102] discovered, as we mentioned earlier, that those companies which lay emphasis on values which go beyond creating profit, are more profitable than those companies which merely formulate their goals in economic terms. This idea lies behind what we today call idea-based leadership. If the company's values are experienced as being important, the experience of meaningfulness is stronger and the workforce is more motivated, indeed even deeply involved in their work. Ethical and moral questions, justice, solidarity and environmental responsibility have become increasingly important in pace with the growing satisfaction of basic human needs.

With meaningfulness as a component of motivation, we have a positive driving force, something much stronger than the psychological reinforcement of desired behaviour. Meaningfulness is also the direct opposite of the teaching approach- nowadays unusual- which seeks to influence people's behaviour by means of sanctions or punishment.

It can, however, be worthwhile to mention the risks which too great an involvement and assumption of responsibility can imply. When management consultants argue for an idea-based or visionary management[103], it has both positive and negative sides to it. Managing people by placing a great deal of emphasis of the creation of meaningfulness, entails that motivation and involvement come from within human beings themselves, and not as an order or command from some superior. However, the disadvantage of this is that the sense of involvement that arises from inside human beings is sometimes so strong that they go beyond the limits of what they can attain in the long term, and run the risk of being affected by exhaustion and burn-out.

There is a long list of further factors which, in differing ways, can contribute to the experience of meaningfulness. In the workplace, it can be colleagues, professional pride, personal development, an attractive physical working environment and other positive everyday experiences which contribute to this feeling.

Hopefully the private sphere of life also provides experiences of meaningfulness. Has this become more difficult in modern or post-modern society? Historically, much time was spent in trying to gather together something to *live off*, whether wages or food. This toil and drudgery made existence meaningful: there were self-evident and obvious motives for the actions which filled the day. Now the majority of human beings in western society have achieved the position where the subsistence problem has been solved: we have to find new sources of meaning. The next stage is to create meaning by having someone or some persons to *live with*. Relationships become intrinsically important: being noticed, being needed and being in demand makes existence meaningful. This aspect is close to the third and last which entails that we experience meaningfulness through having someone or something to *live for*. Thus both people and projects or life goals can serve as meaning-bearers for us. It is indeed quite a wise thought that, as someone has said, quality of life means to have all three: something to live off, people to live with and something to live for.

What effects occur when we experience a low level of meaningfulness or none at all? Above we have mentioned emptiness and listlessness as a destructive state of affairs which renders one

Chapter 6

passive. In a workplace, the experience of meaninglessness brings about a clearly destructive situation. Colleagues who feel this way, have no energy to get involved in their work and, sooner or later, are afflicted probably by poor health. A dialogue in a work group about what is important and meaningful is an excellent activity both for people and for the organisation.

An example of a summary showing SOC at an everyday level and linked to the workplace can be seen in Figure 6.1. This form of listing can provide a basis for analysis, planning and guidance when we are to apply health promotion in a workplace. The list is general in character and applies to some specific workplaces better than others. It is up to the reader to go further on the basis of this and find several factors which can reinforce the sense of coherence in the workplace you personally know.

Workplace Sense of Coherence

Comprehensibility	Manageability	Meaningfulness
Knowledge about	**Resources & support**	**Motivation**
Surrounding world	Material & tools	Visions
Branch	People	Goals
Company's history	Clear organisation	Reasonable wage
Company's organisation	Clear guidelines	Privileges/incentives
Work content		
Working environment	**Possibilities to influence**	**Values**
Own role	Pace of Work	Ethics and Morality
Changes	Planning work	Core values
	Decisions	Just treatment
Feedback from		
Boss	**Competence**	**Positive experiences**
Colleagues at work	Work skills	Relation to colleagues
Customers/clients	Social competence	Relation to management
	Communicate	Pleasant environment
		Humour
	Coping ability	Variation in work
	Physical c.a.	Recreational activities
	Mental c.a.	Self-esteem
	"Distancing"	
	[unwinding from work]	
	Breaks for rest	

FIG. 6.1 *SOC translated to the everyday level at a workplace*

SOC as a temporal dimension

The Lund professor, David Ingvar, was an expert on how the human brain functions. He held that in order to function rationally, human beings must in every situation deal both with their historic experiences and with their intentions and goals with regard to the future. The human brain requires points of orientation in order to make logical and correct decisions. The present is influenced

Chapter 6

by historical experiences and the dimension of the future is the direction which both influences and is influenced by the present.

A company consists of the brains of many people and Ingvar, together with C-G Sandberg[104], put forward the idea of the three temporal dimensions as an important prerequisite in allowing people and companies to function. The "wide awake" or "alert" company is characterised by the wish both to make use of past experiences and to recount its history, while simultaneously it ensures that the vision underlying the work is kept alive. According to Ingvar and Sandberg, it is with the help of these two perspectives, that human beings acquire the optimal preconditions for functioning.

All decisions and actions in an organisation and in a workplace occur in the present. If the enterprise is to function well, decisions must be based on knowledge about, and experience of what has occurred earlier. Decisions must also be in line with the goals and visions pertaining to the future. As a result, every present state has to deal with three temporal dimensions: the past- the present- the future. This temporal axis with its three points of focus can be compared with Antonovsky's SOC. Comprehensibility has an historical starting point, since all the information we have in our brain has gathered there before we can begin to think about it and decide how we are to act on the basis of it. The historical perspective extends from what happened a second ago to what happened earlier in life and perhaps even before that. This temporal dimension is perhaps the one we rank the least important. Or do we carry our historical experiences with us when we meet new challenges? Every day gives new life experiences which can be stored more or less consciously. One way of making experience more accessible and useful is to draw up a "profit and loss account" every evening, making use of such questions as:

- What has this day given me?

- What have I learned today?

Manageability, as a temporal dimension, is that which exists now: it is today, in fact just now, that you can influence your situation. It is presumably an important reminder that it is you who decides and you can do it right this moment. In general there is a choice to be made and a decision to be taken which can influence a situation for the better. It is also possible to wait for a better occasion and do what you thought to do just now at a point in the future.

The Past	**The Present**	**The Future**
Comprehensibility	Manageability	Meaningfulness

FIG. 6.2 *SOC as a temporal axis*

The third dimension, the future is that which lies forward in time. People have to relate their decisions to the goal, the objective, the purpose or vision. Meaningfulness can then arise as an experience that there is something in the future to achieve or bring about. This sounds perhaps unnecessarily philosophical, but it is in fact the case that if we "freeze" a moment in life, there are always three temporal dimensions involved. It is also a way of looking at the sense of coherence and can be a way of going from theory to application.

Workplaces and organised activity depend upon constructive and meaningful actions in order to function and survive. In a workplace where there is little or no contribution to the workforce's sense of coherence, there is a risk of increasing the amount of meaningless activity, of unreflecting work and indeed of destructive behaviour. Experience tells us that when a workforce feels disappointed or abused, the work hitherto directed at collective goals is transformed into activity where people are instead concerned with their own situation and perhaps revenge themselves on their workplace by disloyal or destructive behaviour.

Chapter 6

SOC as a pedagogic model

We can follow Antonovsky's exhortation to use SOC as a model to guide us in the work of health promotion. The theory thus becomes a map which allows us to include important items in our activities to transform health.

It is also important to look on SOC as a teaching model for the work to bring about change and learning processes. How is the work to bring about change to be carried out so that people are involved in it and participate, and where it is effective in achieving its goal? The figure below is an example of the latter way of using the model.

SOC as a teaching model	
Questions Type 1 **Meaningfulness**	What does this give me? Is this important? Is the purpose clear? Do I want to be involved and contribute to its success?
Questions Type 2 **Comprehensibility**	What understanding is needed? What experience is important? What knowledge needs to be extended and updated?
Questions Type 3 **Manageability**	What resources are needed? What conditions / Frameworks are to be applied? How is the work to be directed?

FIG. 6.3 *SOC as teaching model*

If we give an answer to the above questions, we have also contributed to making the work meaningful i.e. the *why* question is clearly answered. The need for comprehensibility figures variously in the form of knowledge, information and understanding.

Manageability also benefits if there are economic, personnel and other available resources. Manageability for the individual employee benefits above all from a management which creates *opportunities* for other people to act, to try things and be allowed to fail.

Since the mid 1990s, SOC has been used as a theoretical model in the school context. The WHO sponsored project, *The Health Promoting School* - at least as far as Sweden is concerned- adopted a salutogenic perspective as a self-evident starting point. A way of creating a school that promotes health is by making the pupils' working environment more comprehensible, manageable and meaningful[105].

> A school which wishes to promote health should, according to this theory strengthen the pupil's sense of coherence. This means that one should strive to ensure that all pupils experience the school as meaningful and have the possibility of understanding and influencing what happens to them there. The development work, broadly speaking, focuses on all activities: teaching, attitudes and behaviour, environment, co-operation both with home and with municipal authorities (p.7)

Both children and adults benefit and feel better from having an existence which provides a good sense of coherence. When we have to learn something new or retrain, when changes are taking place or we are simply confronted with certain tasks, the process becomes easier if the whole thing feels meaningful and we understand, for example, what the change will entail. We are able to cope better if we know the prerequisites and have the level of knowledge that is required. Finally all experience suggests that we learn better and become more involved, if we have the chance to be active and take charge of the situation ourselves. We want to be the active subjects in the situation and to show that we can do it. The management in

Chapter 6

the organisational context which is required, should be supportive and encouraging, not critical and discouraging.

Summary

Our aspiration is to develop health promotion so that it becomes an effective strategy for improving health in working life and in society as a whole. In order to organise and take charge of this development work, the ideas behind it and the goals must be clear. The conceptual starting point of health promotion is salutogenesis and this focus provides a clear direction in the choice of perspective, the fields of knowledge concerned and how we are to work at the practical level.

When our goal is clear, we require a model to guide us, a theory about what is good for health and about what forms the process towards improved health is to take. The Sense of Coherence (SOC) is a very practical and useful model. It has also been shown in many scientific investigations to offer a reliable interpretation of health.

When the model is to be translated into practical activities, both SOC and its three constituent concepts are relatively easy to use. What we have in Antonovsky's SOC theory is originally a general, overarching dimension, a global orientation which can appear abstract or "extremely vague and obscure". This abstract level needs to be clarified and described with additional sub-headings as in Figure 6.1. It is then easier to assign the three concepts a concrete content with respect to a specific setting or context, for example a workplace or school, so that SOC is concretised to an everyday level and thus becomes easy to discuss and understand for each and everyone.

7 Health promotion and its fields of knowledge

The aim of the present chapter is to provide a survey of the theory, in the form of different fields of knowledge, which are involved when health promotion is to be translated into concrete measures and practical health programs in the workplace. With the criteria and prerequisites we have described in previous chapters, it is clear that health promotion is a strategy which covers several fields of knowledge and several scientific domains. The centre of gravity lies in the social and behavioural field, which contains knowledge about humans as social beings placed in the setting in which they live and work. There is also a need for knowledge about technical and other structures created by human beings, which are to be found around us, and about how these should be designed and used in order to promote human health. Knowledge is also needed about those things inside human beings, their mental and spiritual constituents (psychology) as well their somatic, physical components (biology, physiology etc). It can thus be highly positive to have a broad basis of knowledge when we work with health promotion. How is this to be organised? Who can deal with all of this? It is clear that no single individual suffices: it relies instead upon co-operation and teamwork.

In order to apply health promotion as a strategy, there is a need for more professions to become involved. An important role is that played by the person who leads or co-ordinates the process of health promotion. This is a role which requires general knowledge of health promotion and process-oriented health work. A generalist competence or interdisciplinary overview is also necessary in

Chapter 7

order to be able to collect and co-ordinate the combined skills and knowledge of specialists within the different fields, depending on what is required at each moment.

An important knowledge base in health promotion work is also the knowledge possessed by the workforce involved, and their experience of their own workplace and work situation. This is in fact a form of specialist knowledge which no-one outside the workplace can have. According to Sandberg and Targama[106], when we reflect consciously about work processes and work conditions, there is a notable increase in this knowledge. A major point of health promotion is that attention should be paid to this knowledge and that it deserves to be incorporated in analysis, planning and in operational management. It is in this way that the workforce which supplies the necessary energy to the process, are able to participate.

Before we enter into a detailed description of the fields of knowledge associated with health promotion, it is worthwhile to reflect on something which will become relevant when we devote ourselves to a form of health work which is concerned with things on a wider scale. Health promotion involves and influences the whole person and his or her living environment which means that the range of knowledge involved must be similarly wide. What is needed in health promotion is therefore a measure of generalist knowledge, particularly in those people who are to be in charge of this work.

Specialist or generalist competence

Like all other types of work, health work requires competent people. We become competent in what we do through interest, personal aptitude, education, professional experience and so forth. People, who are competent at their job, know what they are about,

Health promotion and its fields of knowledge

they know why things are as they are, and they know how to solve problems or achieve results. How can we produce as many competent employees as possible or how do we ourselves become good at carrying out our work?

Research and the development of knowledge is traditionally geared to specialisation and increasing the amount of knowledge we have about increasingly narrower fields. People who want to be up to date in their own specialism must, due to limitations of time, devote their efforts to a progressively smaller part of the whole. It can be said that our society is filled with more and more specialists who have become increasingly dependent upon each other in order to be in a position to describe and understand things as a whole. As a result, it becomes all the more necessary with co-operation in a team in order to solve tasks and problems in everyday life. Team co-operation in problem solving covers everything from the development of a new technique to the rehabilitation of the long term ill.

Specialists manage rather well on their own when it is a question of carrying out further research and deepening their knowledge. It is in the application to complex situations that the specialist must co-operate with others. Persuading experts to co-operate is not always so easy. Every specialism has its own language and its own way of working. People are anxious to develop and preserve their own domain. This territorial way of thinking can sometimes hinder constructive co-operation. It is an important teaching task to create conditions for co-operation and the collective solution of problems, where everyone involved gains from sharing their knowledge. We usually call this "win-win thinking".

Health promotion requires specialists in a number of areas, but the key function of bringing about smooth co-operation and a well functioning process of development is a leadership task which primarily needs generalist competence. Generalists ought to have general knowledge within the various special fields covered by health promotion, but must also have sufficient knowledge to allow them to appreciate the specialists' competence and to be able to communicate with them. This leadership role involves establishing and building networks, work groups, and teams made up of the appropriate specialists.

It is tempting to think of this generalist- the health promotion's process leader in the organisation- as a teacher, a human resources manager or an organisational consultant who, in addition to being trained in the behavioural sciences, also has knowledge and experience in health work and organisational issues. Here many different backgrounds and combinations can occur, and as usual, it is personal experience and suitability which is crucial for competence. Today, there is an ever increasing group of health educators, health scientists and public health specialists who are receiving education at university level to equip them for this kind of assignment. Such people ought to have knowledge about working life and organisational issues and a deeper knowledge of the teaching methods which are needed in order to be able to lead process-oriented work aiming at change and development.

Delimiting the field of knowledge

We can use the system approach as a help in delimiting what we are to work with and the knowledge we need, in speaking of health promotion in the workplace.

Ohlson and Sandberg[107] have described the bio-psycho-social model in which general systems theory is used to describe how properties and functions at different levels in a biological system are interlinked. An interaction which is necessary for life with the exchange of information and energy occurs between the different levels in a human being's life system. At lower levels, for example at the cell level, it is easier to have an overall picture of the various components involved and the boundaries are clear. The higher the level in the system, the more complexity there is and the functions become more abstract. It is presumably easier to give an account of the physiological processes in a cell than to exactly describe the mental processes which goes on inside a human being's brain or in a collection of human brains. At the organisational and social level, the system consists of many interconnected and interacting subsystems which make it all the more difficult to grasp as a whole and above all to steer.

In this way, we can divide existence into lesser or larger parts of a whole. The whole can then be described as a large system which is composed of smaller systems or subsystems and it is dependent on the latter in order for it to be able to function. When we have to study and describe reality from a system perspective, we can choose to limit our work to a certain level. It is possible, for example, to devote a lifetime of research to how the living cell functions. The knowledge of reality is limited to the cellular level, but the amount of knowledge can still be vast within this particular field.

When our interest is in knowing more about health and how we can work with it in working life , we must draw on a larger part of the whole. That knowledge must extend to the whole human being, the work group and organisation with their respective activities, and perhaps, to some extent, also to society outside the organisation. Would it not suffice to deal with human beings since it is their health that we are concerned with? Obviously both preventive health work and health promotion must cover a human being's life and/or working environment. In order to develop their own health resources, in addition to their own energy, a supportive situation which is positive for health is required. In this way, health in the workplace and health promotion's sphere of work is a matter for the

Chapter 7

organisation as a whole. The organisation can be seen as a system which has to function as well as possible, both from the perspective of the enterprise and the workforce. These various subsystems or domains can be strengthened and improved, like links in a chain. Health work therefore needs knowledge from different fields in order to deal both with the various elements within an organisation and what happens within an organisation.

Knowledge perspectives

When we want to describe a field of knowledge, it is necessary to begin by defining what we mean by knowledge. In order to work with health promotion, various forms of knowledge are required. What type of knowledge and for what purpose? We require both academic scientific knowledge of certain subjects, as well as practical knowledge, depending upon the work and the needs and experiences to be found in what one does. Both perspectives are needed in order for them to be developed in a setting.

Academic knowledge represents the systematic approach, accuracy and critical examination of what is good or bad. It embodies a methodology to evaluate, analyse and develop new knowledge within a field. If we have professional health workers with an academic training, we have a greater chance of coping ourselves with the task of translating the work from the realm of science into practical methods. The flow of knowledge in the reverse direction which occurs when we scientifically scrutinise, document and publish what occurs in practice, is highly valuable for the development of the field.

In order to make this book's account of the field of knowledge more lucid and linked to practical application, we begin it by dividing what we have to say into three different parts or perspectives. In each of these, we can then proceed to slot in traditional academic

Health promotion and its fields of knowledge

disciplines such as teaching methods, psychology, physiology and so forth. This division can also be a suitable starting point when people who work with health and health promotion reflect about the parts of knowledge which are relevant. As noted earlier, it is impossible to be an expert about everything. A certain amount of knowledge about the field as a whole, together with expert knowledge in some specific part of other, is a reasonable and credible level to aim at.

We shall therefore divide our description of the knowledge needed for health promotion in accordance with the following subheadings:

- Knowledge of health and health promotion
- Knowledge of human beings and their life situation
- Knowledge of processes to bring about change

In the work of health promotion, we can imagine that these three parts of knowledge are present in a steering group or similar, which in turn recruits experts for different activities, but also makes use of the organisation's own staff where that is possible.

Knowledge of health and health promotion

Health work involves the three strategies we presented earlier in this book. Health promotion, focusing on a salutogenic approach, needs at the same time to be integrated with both illness prevention and the traditional[108] rehabilitative/curative treatment of ill-health within the framework of the organisation's health work. These three strategies have their own fields of knowledge and their own professional representatives. The professionals belonging to these fields must be able to co-operate and solve problems together and must therefore have a common language. Health promotion, with a salutogenic approach, is largely based on the disciplines of social and behavioural science. Medical knowledge, because of its pathogenic

Chapter 7

starting point, has less importance in this connection. This has been strongly emphasised in the literature,[109] not as a way of calling into question the medical profession in general, but as a counterweight to giving health work a purely medical point of view and slant.

Medicine, represented for example by doctors and nurses, has traditionally played its most important part in the work of prevention and rehabilitation It should, however, be noted that the work of doctors and nurses is being progressively complemented by that of behavioural scientists. When we work with health in an organisation and have to co-ordinate and set priorities between health promotion work, safety work and rehabilitation, medical knowledge has an obvious role to play.

In summary, we can say that a thorough knowledge of the idea of health promotion, its theory and forms of work, is self-evident and important knowledge for anyone claiming this title for their health work. We have to be able to describe this for ourselves and for others. What is characteristic and what conditions have to be fulfilled in order that what we do, can be called 'health promotion'? How does health promotion work alongside and co-operate with other approaches to health? What are the strengths of this particular approach and what problems arise from it? When we know our own field well, we feel secure in discussing and answering people's questions about what we are trying to do and about the aim of our work.

Knowledge of human beings and their living conditions

Health promotion is work to promote wellness in a specific situation or setting, among the people and together with the people who are to be found there. For that reason, knowledge of human beings as physical, psychological and social beings is utterly central.

Health promotion and its fields of knowledge

How do people function and what is it that generally helps to ensure that people's health is improved and maintained? The concrete situation, which this book discusses, is primarily the workplace as an important setting for health promotion work. In order to carry out this health promotion work in working life, in companies, public sector administrations and in particular workplaces, knowledge of these specific settings is essential. What does an organisation consist of and how does it function? What is there in a workplace which influences health and how? Which parts of an enterprise must be involved in discussing health?

ENWHP[110] in the Luxembourg Declaration[111], *Workplace Health Promotion in the European Union* (1997), describes a number of fields which can contribute to health in the workplace. They can be seen as areas for initiatives and as examples of fields of knowledge concerning human beings and the operative conditions in the workplace. This list shows that health promotion presupposes an organisational overview and readiness to apply different kinds of competence depending on what is to be done:

WHP contributes to a wide range of work factors which improve employees' health. These include:

- management principles and methods which recognise that employees are a necessary success factor for the organisation instead of a mere cost factor
- a culture and corresponding leadership principles which include participation of the employees and encourage motivation and responsibility of all employees
- work organisation principles which provide the employees with an appropriate balance between job demands, control over their own work, level of skills and social support

Chapter 7

- a personnel policy which actively incorporates health promotion issues
- an integrated occupational health and safety service

According to this document, health promotion in the workplace should involve:

- strategic management and policy documents
- organisational culture and leadership training
- organisation and staffing
- health promotion activities for personnel
- occupational health and safety(OHS)

We can take this list as our starting point and state that health promotion can involve many questions and issues and as a result requires a corresponding degree of knowledge and skills.

Strategic management and policy documents

In an organisation, there are values and guidelines which are formulated in different ways. Sometimes a management consciously chooses to use central values and ideas formulated as a vision, as a strategy for creating a harmonious atmosphere, involvement and trust in an organisation. This is a form of idea based leadership where good ideas bring people together and help to motivate the workforce as a whole. There are other workplaces where the workforce has not the faintest idea of what its own organisation stands for, or desires. People attend to their customary job without being specially motivated in their work or proud to belong to a certain organisation. Values and guiding principles which contribute to things such as an experience of meaningfulness, pride and a sense of belonging and solidarity, contribute to health. It is an important strategic task for management to make the personnel aware of these positive values.

In larger organisations, in particular, concrete policy documents are drawn up which set out the values and principles the organisation endorses. In small companies, these formulations of policy can be communicated by word of mouth although they would certainly gain in clarity from being set down in print. A well thought out and applied personnel policy is able to discuss many health promotion areas and can be an important support for the operative health work. In formulating health policy, knowledge is needed of all of the health work's strategies and conditions, in order to link these in the policy document to the formulations and systems relating to personnel questions, work and organisation.

Organisational culture and leadership training

The second point in the Luxembourg Declaration deals with the fact that our ways of organising work and the way leadership functions, have an influence on health. Recruiting, training and supporting senior management and supervisors is therefore a self-evident sphere for health promotion measures. Leadership issues and management training is a subject with a long tradition within working life, where every organisation has its own way of dealing with it. Consciousness of, and interest in the importance of leadership varies greatly from workplace to workplace. Smaller organisations which do not possess a special personnel section, is in general worse placed to deal with issues of management recruitment and training in a conscious strategic way.

Most people today would agree that good bosses are important both for the personnel's state of well-being and the enterprise's results. It is, however, a complex and demanding job to be a good boss and as a result, success in this role varies greatly. The management of an organisation has to decide about many questions: what is meant by good leadership? What type of boss and leadership does our organisation require? How do we find the most suitable people

Chapter 7

for these tasks? What conditions make it easier for the senior management to function well in their role? How are senior managers and supervisory staff to be educated and trained? What sort of support do our bosses need? How do we deal with bosses who wish to resign from their job? The list can be extended, but these issues suffice for most people. There are organisations with well thought out programs for management recruitment. There are also still many workplaces which are either unaware of such programs or else are incapable of designing and applying them. This issue is perhaps the really crucial one, if a workplace is to function well and achieve good results.

Competence in these issues is to be found among personal resources managers and human relations specialists. Larger organisations dispose over this competence in their own personnel departments but they nevertheless often choose to supplement this competence by drawing on external resources in the recruitment and training of management and supervisory staff. The majority of investments which organisations make in health promotion today have leadership/management development as a central activity early in the process. The senior management are bearers of the organisational culture and role-models for other personnel in the health promoting work. It can be a matter of the boss's own health or the health of a subordinate or with how the boss wants to contribute to running the health promotion process.

Organisation and staffing

The Luxembourg Declaration takes as its third point, work organisation as an area which is important for health. What is fundamental here is to strive towards a balance between people's capabilities and the demands of their work. Very often it is the staffing problem which prevents a workplace from developing and becoming a healthy workplace. When there is a lack of personnel

Health promotion and its fields of knowledge

to carry out the work required, there is an increase in stress and strain. The work situation can be otherwise fine, but so-called "lean production" in the anorexic organisation wears out people and leads ultimately to the exact opposite of what was intended, namely an ineffective workplace where people become ill.

There is clear evidence that work which allows autonomy and influence is good for health[112]. Maintaining a balance between demands and resources is one aspect of control. It is also a matter of giving human beings the opportunity to influence their own situation at work. Being able to make our own decisions or participating in the decision-making process implies that we feel capable and worthy of respect. Many studies[113] have shown that this is of great significance for health. It makes people feel good and they are more creative, both individually and in a team.

It is important for health that the work is organised and run in such a way that people have conditions which allow them to have the time needed to deal with things, to exercise control by making their own decisions and to have the competence required to do the job. The organisation must also create conditions for a socially supportive atmosphere at work. Once again we are faced with a task that seems impossible. Who can deal with it successfully? Many talented bosses know what to do and certainly succeed very well, while at the same time, there are workplaces where there is a great need for improvements.

Here once again we have an example of the fact that a great deal of work and knowledge is needed to create a healthy workplace. There is need for teamwork among those managers responsible for the health promotion process. It is also necessary to find way of working which will allow us to make use of the combined competence and involvement of the employees in order to carry out the task.

Chapter 7

Health promotion activities for personnel

The fourth point in the Luxembourg declaration covers personnel policy which should be designed so that the health issue and staff health promotion work are given a central role. A very concrete and valuable example in this sphere is the fitness and wellness programs which are to be found in many workplaces. Activities, designed for individuals and groups, aim at making possible and sustaining a sound lifestyle, as well as promoting collective activities and recreation. Since the 1970s, it has become increasingly an accepted part of Swedish working life.

There are great differences between workplaces when it comes to the extent to which employers think that fitness and wellness programs are a profitable investment. In certain quarters, there is investment in ambitious programs and people are even given the possibility of training during working hours. Other employers maintain that questions of personal health and fitness are up to the individual. How are we to prove the economic value of investment in wellness? It involves preventive and health promoting activities which cannot guarantee in advance, a given return for each crown invested. There are, however, many examples of firms who have reported how investment in wellness has led both to increased productivity and a drop in absence due to sickness. In the literature, there is a substantial body of evidence suggesting that there is a clear economic value for the company in having employees who feel well and are fit. There are examples of large companies who have evaluated long term investments of this kind, as well as interesting models of general value.[114]

There are lots of questions about the introduction of fitness and wellness programs which need to be answered. If we have decided to invest in such activities geared to personal development, how do we get everyone to participate? Experience shows that those who are

Health promotion and its fields of knowledge

already physically active become even more active, while those who are couch potatoes, many times continue to be couch potatoes.

In spite of the many hows and whys relating to staff health measures aimed at the individual, there are many examples of success. It is possible to persuade people to be more active, to have fun together and to take more responsibility for their physical capacities in everyday situations. There are no patent solutions: every workplace must design its own program. Once again it is the participation of personnel in deciding what form this work will take, which is crucial for the interest aroused in it.

The surrounding organisation can contribute with a personnel policy, a work situation, social circumstances and practical conditions which will allow people to be active. But in the final analysis, we cannot obtain more physical energy and better health without the individual's own efforts and determination.

At the beginning of the twenty- first century, the tendency in many countries is for more and more employers to invest in fitness and wellness programs. There is obviously an increasing number of employees who participate in activities during working hours or in their leisure time, in order to increase their capacity to cope physically and mentally, to relax or merely to have fun. The last of these is every bit as important as a health promoting factor as any other.

OHS

The fifth issue according to ENWHP[115] and the Luxembourg Declaration is an organisation's investment in well-functioning occupational health and safety service. This, it is held, has an important role to play.

Traditional OHS is aimed at illness-prevention and draws on skills in ergonomics and employee safety at work. There is also a

Chapter 7

tradition of company health care with regular health checks and taking care of accident victims. The medical tradition is obviously a strong one here, because the majority of the personnel involved are physiotherapists, nurses and doctors. Taking into account the spectrum of symptoms which exists today, there is an increased demand for behavioural scientists who can work with the psychosocial work environment in a preventive and actively engaged way. There is also an increased need for health centres to meet the growing demand for fitness and wellness activities within the field of health promotion.

OHS has a long tradition in Sweden and is an established actor in working life as far as health issues are concerned. It is therefore natural that many employers turn to it, both with regard to problems connected with ill-health and their health promotion aspirations. One expects to find in company health care, the professional competence which can offer a complete health package which covers rehabilitation, illness prevention and salutogenic health promotion. The OHS enterprises which are successful in developing skills and competence, and which also are successful commercially, will continue to be important actors in the future. If doubts arise about the ability of those actors, on the basis of current needs, to deliver health, then other actors will take over. There are already a number of examples of this happening.

Knowledge of processes to bring about change

What do we have to do to get the health-promotion work going and how can we steer it in order to achieve the best results?

Achieving results in this context can only be done by bringing about change. Something has to happen between point A and point B which entails that point B is seen as a better state than A. When we think in this way, we challenge everything that constitutes human beings' current reality and truth. It can create a sense of insecurity

Health promotion and its fields of knowledge

to speak about change since the people who are exposed to change know what they have, but not what the proposed change will bring about. It is not easy to describe a state of affairs which does not exist, in a way that makes it convincing and attractive. Someone can think that the change sounds too much of an upheaval, indeed somewhat threatening. We should perhaps use the word development instead, which implies that much of the original is preserved and merely undergoes certain improvements.

When we speak of health promotion as a process of change, it is a question of work that continues over a certain period of time, which has a clear goal and which encourages participation. The odd lecture on stress or some other topic does not meet the requirements. Similarly an hour's massage at work does not create change, even if it may ease cramped muscles. These various activities generated by the great interest in health, are intrinsically valuable, but they are of even greater value when they are placed within a setting. Health promotion presupposes that we begin with the general goal of improving health in the workplace, linked to the organisation's overall mission. It is then a question of creating an organisational framework for health measures, so that all promising suggestions about activities can play a constructive role. In order to succeed with the work of health promotion, there has to be a legitimate goal, a structure for carrying it out, arrangements for the participation of interested parties and someone to take charge of the process.

In order to get to the nitty-gritty and illustrate the competence needed in the workplace health promotion (WHP) process, we can make use of the training manual[116] brought out by the European Foundation after its study[117] of health work in 1400 companies. In the analysis of the results of this investigation, it was found that in many places, there was a lack of the knowledge and skills which are needed for health promotion in the workplace. As a result, an inventory was made of the courses in health promotion available in Europe. It was discovered that there was a lack of suitably

Chapter 7

designed training in this form of process oriented health work in organisations. It was decided to produce a training manual which took as its starting point what was called an idealised health promotion process.

According to the manual, the idealised health promotion process consists of seven steps:

- Marketing health promotion
- Setting up structures
- Assessing needs
- Developing a plan
- Implementing the plan
- Evaluating the initiative
- Amending the plan

The training manual states yet again that there are several professional groups who need to be involved in the process. Since there is no self-evident professional category which can assume the main responsibility for the process, one chooses instead to describe a number of functional roles or actors who have the task of seeing that the different steps in the process are carried out properly.

A set of six roles which need to be fulfilled for successful health promotion:

- The expert
- The advocate
- The deliverer
- The participant
- The change facilitator
- The decision maker

Health promotion and its fields of knowledge

The expert is the person who supports or leads the process and who ought to have knowledge of health promotion, research methodology, data analysis and work safety.

The advocate has sufficient knowledge of health promotion and marketing (communication) in order to be able to present and carry on a dialogue in the field with various target groups.

The deliverer is the person who carries out the concrete activities. This is a matter of giving training courses, making contacts with internal and external resource people, leading groups and dealing with anything that can occur.

The participant has, unlike the case in other forms of health work, a participating role in workplace health promotion. That means that every participant/employee has a responsibility and is expected to contribute actively in the implementation of the health promotion process.

The change facilitator has an important function since the health promotion process implies changes both for individuals, for the working environment and the organisation. The person in question must be able to deal with group processes, organisational development and management problems in general. The skills required for this post are most often to be found in corporate management or among organisational consultants.

The decision-maker is often a senior manager who assumes the responsibility for ensuring that the health promotion process is integrated in the organisation's general strategy, and who makes decisions about the allocation of resources for health promotion activities.

This list of roles ought to be seen as a fundamental model of how different types of competence can contribute to the different steps in the health promotion process. It can serve as a guide when we want to design training for those who are to work with health promotion

Chapter 7

in organisations. When we come to actual practice, it is more or less valid, depending on the particular circumstances. Different parts differ in importance, depending on the kind of organisation we find ourselves in: the company's size, the educational level of its personnel and their experience in participating in development work, the level of aspiration when it come to health promotion work, the supply of resources in the form of human beings, time and money etc -all these things have an effect. In every specific situation, the existing conditions must be dealt with in an optimal way.

The European Foundation's specification of educational training ends with a crucial remark in which it is emphasised that it is the participants who are the most important driving force in workplace health promotion. Experts are needed, not to direct the process, but to support it:

> Secondly in identifying the roles needed for successful WHP, the specification emphasises the fact that good WHP involves some level of organisational change, and that the participant is a legitimate and important driving force for the entire process. In addition, it recognises that the process is not expert led, but is supported by expertise.[118] (p.28)

Health promotion is about change. If we are to bring about healthy workplaces, this means for most workplaces, new ideas and a new way of working with health issues. It is therefore important to understand the character and conditions of the process of change: what it means for the individual, the group and the whole organisation to initiate and carry through processes of change and development. How do we create the determination to bring about change and how do we deal with lack of interest and opposition in a way that will benefit the overall goal? From the moment an

Health promotion and its fields of knowledge

initiative is born or a decision is taken, appropriate actions with clear goals have to be carried out.

The extent and difficulty of the process of change can vary greatly. Sometimes we implement simple, minor initiatives for health in the workplace which are easy to describe and organise. An example is starting a study circle, dealing with one's ability to cope in everyday life, which is scheduled to meet ten times in the course of a term. Here there is not much that can go wrong; the emphasis is laid on the actual implementation itself and it is easy for people to have an overall view of the subject and what it is about. Provided we have a good person to lead the circle, the right study material and distribute information in plenty of time, the results should be good.

It becomes more difficult when we have large projects which have to be prepared in advance, planned and carried out throughout the whole organisation. In this case, great demands are placed on knowledge and experience of processes of change in organisations. The organisation is a jungle and pitfalls abound. In this particular setting, knowledge of change and the theory of change form a large and important area of knowledge. It can be studied to some extent, but it is much more a question of on-the-job learning by experience and of participating in the chaotic open systems which organisations often resemble.

Health promotion's fields of knowledge are to be found in reality and it is there that we have the concrete experience of how things work in practice. However, in order to be a clever and successful practitioner in a certain field, theoretical knowledge is also required. As has been described at the beginning of this chapter, the situation which confronts us, has to be viewed with a critical, more theoretical eye. It is rather like viewing things from a helicopter, or taking a step to one side in order to reflect over and analyse the problems

Chapter 7

with which we are faced. Knowledge which is based on academic research also helps us to keep a good grip on theory and practice and assists us in checking and evaluating the practical results.

Academic knowledge

In our society, it is the universities and other research institutions which are responsible for producing new knowledge based on research and for disseminating this knowledge in society via educational courses. Health promotion as an area of knowledge is also the subject of discussion and research. New knowledge is published in books and scientific journals. Because health promotion is a very broad field of knowledge, there are many traditional academic disciplines or subjects which can contribute knowledge to it. Health promotion has its roots in public health and public health work and it is in accordance with this tradition that Bunton and MacDonald[119] have proposed various areas of knowledge for training experts in health promotion.

This list of relevant disciplines has served as a guide for many educational courses and underlines, by placing its centre of gravity in the social and behavioural sciences, the subsidiary role of medicine in this form of health work. The authors distinguish between what they call *primary feeder disciplines* and *secondary feeder disciplines*. The primary disciplines are psychology, sociology, teaching methods and epidemiology.[120] Among the secondary disciplines are, for example, economics, marketing, communication and philosophy. The majority of these subjects form part of academic courses geared to public health science and health education. When we focus on health promotion in working life, we may state that the subjects listed are more or less central and if anything is to be called the principal subject, it is teaching methods (pedagogy). It deals with learning and learning is development and change and therefore fits in with health promotion's agenda rather well.

Health promotion and its fields of knowledge

We shall content ourselves with briefly commenting on the so-called primary subjects, but making an exception for philosophy which we shall include since it plays an increasingly important role in health promotion work. Another addition to Bunton & McDonald's list is system theory, which is in fact not an academic subject, but rather a more general theoretical principle. It is included in this section because it has much to give both in organisational thinking and in health promotion.

Philosophy

Philosophy is one of the first and oldest academic disciplines. During ancient times and the Middle Ages, the term philosophy was applied to all systematic thinking outside the field of religion.

When science seeks new ways among the established disciplines, philosophy ought to rank high in the list as a help in reflecting and critically assessing the utility of new knowledge and new methodology. Philosophy supplies principles for how new theory, new knowledge and new concepts can be formulated and for how these correspond to the reality we live in.

Moral philosophy (or 'practical philosophy' as it is called in Sweden) is concerned with ideas about human beings' moral and social existence and the fact that we all the time make evaluations and decisions about how we ought to act. The concepts of ethics and morals have in recent years attracted increasing attention. There is trend towards writing and lecturing about their importance and this is noticeable, not least in working life. It is perhaps only a transient phase or else there is a deeper reason in the post-modern transformation of society which for a time allowed norms and values to become eclipsed. In 1989 already, Koskinen[121] asserted that interest in value questions can be linked to an increasing rootlessness in our society. He writes about the unitary society,

Chapter 7

with a monarch, a nation and a belief which has become pluralistic with all the difficulties that are entailed in finding a functioning system of norms. Another explanation, according to Koskinen, can be that people have become satiated by consumer society. As a result, questions about a more profound notion of quality of life are instead given greater prominence.

This section does not aim to discuss philosophy at depth or to explain what the concepts of ethics and morality mean. We can simply sum up by stating that these are present when we speak about values and their foundation. They recur in several ways when we devote ourselves to health work in organisations. For example, are the organisation's fundamental values discernible in its policy documents and are they given priority? To what extent can the organisation summarise or formulate its view on issues of values? Policy documents vary in importance in different types of work. They can contain formulations which provide preconditions and support for the work of health promotion. An important question in this context is how and by whom the strategies described in these documents are formulated. The philosophical issues are tied to human beings, cultures, religions etc. The discussion about what constitutes good ethics and morals is therefore best conducted by bearing in mind the real conditions which currently apply.

When we describe and train people in leadership, according to Trollestad[122], value questions are crucially important. The leadership we strive after follows from some form of ideology, whether conscious or unconscious. In this context, we can discern the results of the morality we embrace and the view we have of organisation and human beings. We want to have senior management who act in a way which is consonant with the core values of the organisation and can be considered morally acceptable. We want to have an organisation which is effective, but which simultaneously protects people's rights and liberties.

Health promotion and its fields of knowledge

Every work group has also a norm system. This is often implicit rather than explicit, but with a secure foundation among those who have been there some time. Is there a reason for discussing value issues in professional life and in a work group? What values and principles are important in the health work of the organisation and in the health project under discussion?

Ethics and moral issues can be given concrete form in certain everyday questions:

- What should I, as the person responsible for the activities, do in this situation?
- How should I implement it?
- What is the goal? What is it meant to lead to?
- Is the desired result good? Good for whom?
- Can my decision have certain undesirable consequences?

We always have some normative foundation as our starting point for thinking and acting. Philosophy helps us to deal with these sorts of questions. The normative foundation in working life, in leadership and in health work becomes particularly apparent under headings such as ideas, viewpoints, policy, priorities, motives, motivation, driving forces and meaning.

This is a situation where we often have to compare/evaluate different formulations, decisions and alternative courses of action. We assign different values to the alternatives we have to choose between, and this then decides their relative priority. Sometimes it is easy to make up our minds. We work for the common good and accordingly our decisions should reflect this. On other occasions, we end up with a conflict of interests where the decision benefits one party, but is to another party's disadvantage. "It is good for the patients but less good for the personnel and it is not at all good for the economy". On many occasions, it is a question of "choosing between the devil and the deep blue sea".

There are many more examples and reason for giving philosophy and value questions a larger role in a field of knowledge such as health promotion. There is no space to discuss this in more detail here, but philosophy ought to have a more evident position in longer educational courses dealing with the field.

Psychology

Health promotion aims among other things at creating a balance between the individuals' potential and the demands which are made upon them. Individual potential consist of both physical and mental resources. The demands can arise from within in the form of thoughts we ourselves have, or from without- that is to say from people in our surroundings. It is human beings' thoughts and their response to the demands made upon them, which largely decide if they will be able to cope and how they feel.

Much of what we experience has to do with relations to other people. The social situation involves strains, demands and stress. It is also the social situation which supplies us with resources and support in order to function successfully and to cope. The subject of psychology has therefore a prominent role in health promotion. Psychology studies peoples' behaviour and mental processes, both as individuals and in groups. How do thinking, emotions and co-operation function in a group? These processes concern cognition, emotion, communication and relationships.

As a subject and area of research, psychology is still relatively young, but it has nonetheless been divided into subsidiary specialities and found many applications in different settings. Psychologists have reflected about the psychology of personality, the psychology of development and social psychology for some time now. Applied psychologists have defined such areas as health psychology, work group psychology, organisational psychology, the psychology of sport, the psychology of leadership etc.

In the traditional literature within Health Promotion[123] psychological theories are often used to explain behaviour and behavioural change in the case of behavioural habits and healthy lifestyle. It is in Health Education, above all in the Anglo-Saxon part of the world that new theories about behavioural change are developed. A model which is often cited, the so-called *Health Belief Model*, used by Prochaska and DiClemente in 1983[124] to describe the various stages we go through before we have firmly established a new habit.

It can also be used to describe the phase in which people or groups undergoing a process of change find themselves, and therefore give support to suitable initiatives to facilitate the process of change. It is important to know the target group or human being in order to communicate the message about a change in health-related behaviour or life-style in the right way. The different stages of change according to the Health Belief Model, as Nutbeam and Harris[125] describe them in the figure 7.1, can be applied, among other things, to giving up smoking, weight programmes and to physical activity.

Chapter 7

Stage in the process of change	Response / activity
1. Pre-contemplation	Sees no advantage in altering one's behavioural habits. The disadvantages weigh more heavily. Resists social pressure to change one's ways.
2. Contemplation	Intends to change within a certain time. Can see certain advantages with changing one's habits.
3. Preparation	Has decided and has begun to prepare for changing. Finds out what one should do, gathers information, literature, attends courses etc.
4. Action	Has a plan and begins to follow it. Great risk of 'return' to old habits.
5. Maintenance	The new habit has been functioning for sometime and is progressively becoming a fixed habit.

FIG. 7.1 *The transtheoretical (stages of change) model*

This model can be seen as an example of how theory and practice working together can develop and improve the work routines we use. Through greater understanding of how we human beings think and function, health educators are in a better position to choose how they should communicate with their target group and what information they should make use of.

Psychology is an important fundamental field of knowledge not only in health education, but also more generally in health promotion work which aims at bringing about a change in a workplace or organisation. For reasons of space, however, this is not the place to go into further details about the field of organisational psychology.

Teaching methods (pedagogy)

The bringing up of human beings, education and learning are examples of activities in society which pedagogy seeks to describe and explain. Based on politics and a view about the society we wish to create, teaching methods are devised and tried out on every new generation. The knowledge which the new generation needs is defined and administered by the adults or the decision-makers in society. Pedagogy thus contains both studies of what should be learned or changed and how this is to be done.

The role of the learner is central to pedagogy. What is the relative importance of the person who has the knowledge and who has the task of communicating it, compared with the person who wishes to acquire the knowledge? Historically the teacher has most often been centre-stage and the learning situation has been authoritarian in character. It only more recently that people have realised that in fact it is the learning side of the equation- that is to say, the pupils and their interests and questions -which is of chief importance in learning. The view of human beings has altered so that we see them as active and creative individuals seeking to realise their potential, and with wills and aspirations of their own. The result has been that we witness an increasing number of examples of teaching methods which are based on this outlook. This view of human beings is also basic to the current processes of change in the case of organisations and serves as one of the main principles in health promotion.

Pedagogy is a practical and very useful area of knowledge and pedagogical knowledge and experience is of great help -indeed necessary -in order to carry out the work of change and development. There is always an element of learning involved in these things and there is therefore reason to raise pedagogical questions such as:

- What is knowledge, what is change?
- What knowledge? What must we know?

Chapter 7

- How is the learning / change to take place?
- What roles should the different actors have?
- How are we to evaluate the results?

Sociology

For anyone working with people in organisations or in a society, sociology is of great importance. At the macrolevel (=structural level), sociology seeks to explain how society functions by studying everyday activities such as working life, religion, economics and education. These social functions are organised and institutionalised in e.g. industries, schools, families, voluntary associations and politics. Sociology asks what sort of people are to be found in these. What purpose do they serve? How do they work/function? What normative values are customary? How are the activities financed? Who holds the power and runs things?

These and many more questions form the starting point when sociology helps us to understand how a company or a public authority functions. This knowledge of the bigger picture lessens the risk that we go astray or become incapable of action in confronting a system where we do not understand how it functions or what is happening.

Sociology also works at a lower level, at the so-called microlevel, where human behaviour is studied. What do human beings do and why do they act or behave the way they do in these organisations and institutions?

According to Nicki Thorogood[126], it is arguable that the heart of sociological research lies in combining the questions at both these two levels. In working life, for example, it is interesting to know what effect the form of an organisation has and how the way power is distributed, influences the way people think and act. Given this perspective, the way we organise and lead a company, determines how people behave. Another perspective is to reflect about how people's combined action influences the organisation and thereby decides what sort of workplace it is. According to Thorogood, it is important to realise that this is, in practice, a dynamic process in both directions and to bear this in mind in the work of health promotion where we are involved in trying to promote change at the organisational/structural level as well as at the group and individual levels.

Health promotion is an approach to health work which has begun to establish itself in society and which *inter alia* is applied in organisations. It is also interesting to study how this establishment of a new way of looking at things and of working, takes place. In order to discover the role this particular area of interest has in the larger picture, sociologists[127] have examined the nature of health promotion, and its origins and the form it takes. What of the wider context? How does health promotion tie up with, and interact with, or influence other forms of health work and "movements" in time? This is a form of background knowledge which ensures that those working practically with health promotion have a foundation to stand on and can see their own work as part of a larger context. It then becomes easier both to follow - and contribute to – developments in this field.

System Theory

Sociology is an example where the choice can be between seeking explanations of human behaviour only in the individual or in the structure/organisation. An alternative is to apply a system point of view and study the significance of each part of the system and how the different parts influence one another. System theory is to be found in various subject areas such biology, physics and behavioural science.

Chapter 7

In biology, one looks upon nature as a whole, an ecological system where there are system levels ranging from the single cell to the whole of the biosphere and even farther. An organism somewhere in the system is not only dependent on organisms of its own kind but is influenced by the system levels which are above and below.

In physics, system thinking is applied in thermodynamics which describes the behaviour of gases and liquids. Cybernetics, the theory of control and communications in technical systems, had its breakthrough in the 1940s and since then has had great importance for the development of computer technology.

Since the 1960s, the system approach has been used much more in behavioural science. It helps us to describe and understand the organisation. It helps us to work with change. System thinking, and its application in *theory of logical types,* has acquired importance for communication and conflict resolution.

System theory or the system conceptual approach is not an academic discipline in itself, but rather a theoretical foundation for both thinking and practical work which can be used in health promotion. Here is a brief historical sketch with some examples to show what is involved and to arouse interest in reading further in the reference literature.

The anthropologist and psychiatrist Gregory Bateson[128] holds that we human beings, particularly in the Western World, have great difficulty in assimilating new viewpoints which break with tradition, reason and normal logic. We are dependent on "linear thinking" where cause and effect must be visible, logical and predictable. We have less room for spontaneity, intuition and emotional logic which does not fit into a technological- economic society like ours. Bateson holds that we need to incorporate our thinking and striving for development within a holistic way of thinking in order to secure our future survival.

Among the various aspects and applications of system theory, the following may be noted:

More chaos and less linear thinking. A system is characterised more by diversity and complexity than by a simple connection between cause and effect. Linear thinking, the idea that because B follows A before it must do so again, is not always helpful. A system is not so predictable. There is a series of factors which can affect the result and which we cannot control or even know about. If we look upon the organisation as a system, it is then perhaps easier to put up with the chaos there is, and to learn to act in accordance with the conditions that chaos entails.

The system can be self-regulating. A system can incorporate self-regulating rules in order to keep the system more or less in balance by feed-back (homeostasis). This provides a kind of system equilibrium within a certain range of values. The normal state in a system is disorder (heterostasis, chaos) but provided the disturbances are not too large, the system can hover around its "normal value". The regulation of human body temperature is a good example of this.

Open systems. In order to survive and continue to develop, new energy must be supplied to the system. In a closed system, energy gradually seeps away (entropy), a process of dissolution which begins directly the system is formed. In an organisation, one can speak of mental entropy as the result of unclear goals or bad leadership. The open system is characterised by activity, dynamics and continual change in order to adjust itself to new conditions in the world. Bo Ahrenfelt[129] strikingly describes how the workplace and organisation can be interpreted on the basis of this way of looking at things.

Chapter 7

Problem-solving. System-thinking helps with the solution of problems and with bringing about change. Normally we solve problems in the way we have employed in the past and using the methods which are familiar to us. The problems can, for example, be conflicts which on occasions can be difficult to resolve and which defy continual efforts to find a lasting solution. In order to solve the problem and to communicate effectively, according to Watzlawick[130], we have to distinguish between various logical levels involved in the problem and in communicating.

The solution to a conflict can, at one level, mean two people arguing more often and more effectively. The same conflict can also be resolved at a higher logical level where the technique can imply that the parties to the dispute discuss instead what steps they should take to deal with their conflicts. This change of system level can be seen as a radically different strategy for finding a solution. In the literature, this approach is called a *second-order solution*.

Epidemiology

When we choose to place epidemiology among the primary areas of knowledge (primary feeders) it is because this discipline provides us with a tool to describe the picture of illness in e.g. a country or a large company. This is useful for evaluating the effect of health initiatives which have been implemented. Epidemiological data can also provide the basis for decision in settling priorities between different population groups or different illnesses. Public health is traditionally concerned with preventing ill-health and from this point of view, knowledge about which illnesses occur is self-evident and important.

Health promotion and its fields of knowledge

Several subject areas- old and new

There are other fields of study which are useful in health promotion and which can be called academic disciplines. Anyone concerned with plans for physical activity and lifestyle issues needs to have an insight into anatomy and physiology. Anyone engaged in planning, directing and evaluating requires knowledge of economics, the methodology of evaluation etc. However, we shall not discuss any further subjects here. Health promotion is based on teamwork and co-operation between generalists and specialists. When we are preoccupied with a concrete task, it is a matter of recruiting the competence which the respective activity and stage in the process requires.

It may be observed that the great social interest in health issues ensures that new perspectives and fields of research appear on the scene. Today researchers seek to know more about the organisation of work and health, economics and health, design and health, to name only a few things. We are presumably only at the beginning of a growing interest and an expansion of knowledge linked to health issues. This development is steered by the contemporary problems of ill-health and the economic driving force behind such an issue. The traditional medical professions, with their predominant influence will eventually be forced to concede a greater role to economists, technologists and social scientists who wish to participate in explaining how people can remain healthy. Those who come after will be able to see how far the ideas and proposals in the current literature are subsequently developed.

Summary

The need for knowledge in health promotion is based on a generalist way of thinking. We have to include human beings in all their aspects and in their living environment. Our starting point is

Chapter 7

therefore a general one when it is a question of describing the field of knowledge concerned.

We have given an account of the field in two ways: first of all, on the basis of three perspectives which are linked to what is needed at a practical level to run the health promotion process. Among the desiderata are (1) knowledge of this particular approach to health work and moreover (2) a broad knowledge of human beings and their living environments, as well as (3) knowledge of the process of change. These three dimensions of knowledge depend on each other, in order to be able to deal with the work of health promotion.

We have also gone through a number of traditional academic fields of knowledge and theory which are important for health promotion. These are of value for anyone wishing to deepen their own knowledge, or for those who are involved in educating those who are going to work with health promotion. Scientific knowledge and the scientific approach are necessary so that what is a new field of activity will not find itself groping blindly or going astray.

With this, we leave the part of the book dealing with health promotion's theory in various senses. In what follows, we shall devote ourselves to the practice of health promotion where we shall describe in a more concrete way the prerequisites, conditions and approach associated with this form of health work. How do we get things started? How do we prepare our plan and program for promoting health in the workplace?

In entering this discussion, we should remind ourselves that it is very practical to have a good theory. The strength of the concept we call health promotion arises from a clear link connecting idea to theory and practice. Practice should thus be guided by the idea of salutogenesis and shaped with the help of the Theory of Sense of Coherence (SOC) and associated fields of knowledge.

8 Health promotion in practice

Promoting health and drawing attention to salutogenic factors can be done in various ways. Health promotion is a way, or rather a strategy for health work. By strategy, we mean a longsighted overarching approach which has a clear conceptual starting point or aim. This conceptual starting point means that we have assistance in determining what the health promotion work should involve and what knowledge is relevant for the health worker pursuing this particular approach. The idea that forms our basis is *salutogenesis*. We have presented this earlier and it can be expressed as a question: what helps us to stay healthy or improve our health?

The international tradition of health promotion has been criticised for wanting, more or less, to swallow up everything that has to do with health. Promoting health is accomplished by all types of health worker and by means of every conceivable method. According to David Seedhouse[131] this pragmatic approach has meant that too little time has been devoted to reflecting about fundamental values and the definitions which determine the view of health adopted, the knowledge involved and the choice of method.

Since the interest in wellness programs and health promotion is considerable, there are also many proposals for how this work should be carried out. In the international literature, there are examples both of principles which help to guide us in concrete matters and complicated theoretical models which aim at describing what one does and how health promotion is to be pursued. Europe and USA are responsible for most of the literature on health promotion which is published, and in more recent years this has included literature dealing with *Workplace Health Promotion (WHP)*. Descriptions of

Chapter 8

projects, programs and evaluations are available in great numbers via the Internet for anyone wishing to delve more deeply after reading this book.

In the material which is published internationally, there are many different models which describe health promotion and this often has the result that rehabilitation, prevention and promotion of health are bundled together under the same heading which leads to a lack of clarity about whether these derive from an illness or health perspective. Seedhouse argues that this confusion of models is a sign of a lack of criticism, and this can be partly due to the fact that health promotion, in comparison with other disciplines, is young tradition which has not succeeded in formulating its fundamental idea and theory. He writes:

> Health Promotion theorists get in a muddle about models because health promotion has not yet developed a tradition of critical analysis. Unlike the theoretical writings to be found in more established disciplines, those who write about health promotion tend to be concerned much more with getting a message across than with establishing precise meanings and applicable theories. (p.44)

In the practical situation where health promotion is linked to a setting, three strategies are necessary and both perspectives- illness and health- must play their part. For example, it is the needs and conditions of the actual workplace and organisation which decide which strategy or strategies should be used. However, it makes things easier if we distinguish between perspectives and try to be crystal clear in our use of concepts so that competence, ways of working and evaluation methods can be applied and co-ordinated more appropriately in keeping with the objective.

Health promotion in practice

If the task or goal is to prevent illness, the efforts should be as effective as possible and correspondingly adapted with a view to attaining this end. The evaluation of results must also be carried out in a way which suits the method adopted and the stated goal. The conditions vary greatly depending on whether we are evaluating the results of rehabilitation, the elimination of risk factors or measures of salutogenic health promotion.

When the task is that of improving the conditions for health among the workforce, the direction, the formulation of the goal, the choice of method and the evaluation procedure must be adapted accordingly and there must be internal agreement about this. It can be asked, for example, if it is wise to measure changes in ill-health when we are evaluating the effects of efforts directed at supporting and improving health or the prerequisites for health. Often through force of habit, we continue to measures things which are not directly relevant to our task.

There are many other examples which lend support to the idea that in health promotion work we need greater critical analysis of what we are doing. We also need, as Seedhouse [132] asserts, a unifying goal and a theoretical foundation or model which can guide our work..

In this book, we present a model where the idea of salutogenesis serves as our guide in developing health promotion in working life.

It is easier to link ideas and theory to practice, when we are dealing with a concrete setting like a workplace. Moreover, it is also the setting where the author has gained most of his practical experience.

We look upon health promotion as a strategy for wellness-oriented work where we have:

1. A conceptual starting point (salutogenesis) which provides the direction our work is to take and which determines or guides the analysis of what we mean by health, and which view of health and concept of health are potentially useful. This idea also guides the choice of fields of knowledge, the way of working and assessment procedures.
2. A formulated theory (SOC), in line with the idea, which guides us in determining what form the conditions which help to develop and maintain health, are to take.
3. A model which specifies how the change which will lead to the a state of affairs which is positive for health, is to be brought about

Why health promotion?

First of all, we start with a reflection about the motives for investing in health promotion in an organisation. According to our reasoning about the different strategies, the motive can be to resolve problems which have already arisen, to prevent problems from arising or to develop resources. The major survey[133] which the European Foundation published in 1992, investigated the reasons why companies were prepared to invest in health work. Among the 1400 companies in the survey, more than 60% said that their most important reason was linked to legislation, the health problem in a working environment, psychosocial problems or productivity problems. It was thus mainly a matter of health work which aimed at solving problems within the organisation. In the survey, there were also companies who had other reasons which were more connected to development such as e.g. creating a safe environment and promoting a healthy life style among their employees. A small

Health promotion in practice

minority were more concerned with making organisational changes in order to improve the health of their employees.

Since the publication of this study, interest in health promotion has grown in strength among those concerned with health work in working life. The arguments used for justifying this health work have increasingly stressed what is positive from the company or organisational point of view e.g. factors such as increased productivity, reduced absence, an improved atmosphere at work, a more attractive workplace (corporate image), good recruitment and less change of staff. The importance of health is advocated by private consultancy firms, national authorities[134] and WHO[135] as international actor. The European tradition appears to be a step behind its American counterpart which at an early stage began with "Health and Productivity Management"[136] in order to create better economic results for corporations. American company "fitness programs" and other activities were shown to be profitable, principally by reducing the insurance costs the companies there pay for many of their employees. Today, the health worker promises, in addition, increased productivity if the company invests in "Health Promotion programs".

This argument, with its economic appeal to the organisation's advantage, has an increasing role, when compared with the discussion about the health work's value for the individual. This reflects the current fundamental value judgement whereby we always seem to be dependent on economic arguments when it comes down to our priorities and investments. It is not enough merely to have an empirically based logical understanding of the connection between the well-being of employees and economic results achieved by the enterprise. Through the years, many attempts have been made to calculate how profitable investment in health is. Such "cost-benefit"[137] calculations are in general somewhat vague and function mostly as a conceptual model for linking health to increased productivity.

Chapter 8

Well-being and good health ought to be regarded primarily as something of intrinsic value for the individual and ought to be an obvious priority from a humanistic point of view. An investment in the health of employees ought to be based, not on some profitability calculation, but on the fact that the company in question has, for example, formulated certain central value judgements and policies which emphasise human values. No doubt this can be viewed by some as being of less value in purely commercial terms, but as we have indicated earlier, it has been shown that such values and priorities can pay handsomely in terms of company performance and sustainability.[138]

Irrespective of the position we start with, it appears to be absolutely clear that the individual person, organisation, employer and society at large, gain from the state of health being good. Given the ever-higher costs entailed by illness today, the figures increasingly speak in favour of health promotion work. More and more organisations also see the possibility of improving and protecting their productivity and competitiveness by creating better conditions for human beings.

Increased productivity and human sustainability as a result of investment in better health conditions is a development subject which certainly both researchers and practitioners will devote great attention to in the future. The professional actors and scientific disciplines which wish to participate in giving an answer to the mystery of health will also grow in number. From the beginning it was a medical question which concerned solving the problem of human illness at the cell or organ level. Afterwards the social and behavioural scientist entered the picture and extended the health perspective to cover human beings and their living environment in its entirety. Today researchers in organisation and management theory have realised that human health is pre-eminently a structural problem which has to be considered in all organisational and

Health promotion in practice

management work involving human beings. Company senior management must reflect on how the organisation should act in order to be sustainable and survive in the future.[139]

The discussion of illness and health in working life has evolved into an organisational issue- a matter that is, in the highest degree, strategic in character. Increasing absence from work due to illness is a cost and a driving force which is increasingly becoming something to be reckoned with by the majority of employers. Many organisations, therefore, set out to reduce absence due to illness. A number turn things round the other way and reckon out how many employees are "long term healthy"[140], a way of calculating which can be stimulating and which can encourage increased workplace attendance. However to calculate and draw attention to the proportion of long-term healthy also entails a risk that the number of people who are in fact ill, but are nevertheless at work, increases. Being long-term healthy means that one sometimes goes to work despite being somatically ill and in fact should have registered as being sick. This hidden number of people who are ill, but who still continue to go to work, is fairly widespread depending on the type of work and the conditions which apply to it. People ask themselves: "who is adversely affected if I do not go to work today?" People, who know that their colleagues, pupils, patients etc. will have a difficult time when they are off, are probably less prone to register as sick. On the other hand, it can also be so that the workplace is, comparatively speaking, a "healing" environment where we recover our health more rapidly than by lying at home in bed. People have to decide for themselves, what is the correct alternative in the light of their own circumstances and situation.

Put somewhat simply, we can say that today it is said that there are three different reasons for being interested in the health issue in working life. The most common one in Swedish organisations is to reduce absence due to illness. A more pro-active goal is to

Chapter 8

increase the number of people who are long term healthy. The third reason, which is growing in strength, is to maintain and improve the productivity of the workplace as a result of good working conditions and good health on the part of the employees.

Irrespective of the underlying reason or goal, researchers and company leaders perceive that health work must be incorporated in several areas such the organisation's policy, its culture and its basic set of values, in issues dealing with competence and staffing, organisation, management and personnel development. As a result, the way of discussing and dealing with health questions must be broadened. Health in the workplace is a joint task which is line with modern leadership and organisational thinking. "Transformative leadership", the "learning organisation" and an evolutionary approach to change, are some examples of traditions of thought[141] which can also be viewed from a health perspective.

This has made health promotion an interesting and important strategy where employers together with employees, occupational health and safety staff, and other actors, develop forms of work which are much less concerned with just reacting to ill-health, but instead aim at improving health at the workplace. When the objective is a low rate of absence due to illness or a high rate of attendance at work, we measure success in terms of how many employees are at their work. It is an important and profitable goal which can help to reduce health costs. Moreover, the employee who is at work helps to attain the goal of the enterprise concerned. Productivity, however, varies from day to day and from individual to individual, and also depends on how employees view their work situation.

By investing in health promotion, the goal can be set higher than merely reckoning how many employees are present at their work. The aspiration to create better opportunities for health, does not lead merely to increased well-being. There is a high probability

Health promotion in practice

that productivity will also improve when well-being and pleasure in one's work increases. The model of health promotion we propose, means bringing about a process where the goal can be described in figure 8.1. Many workplaces have a low rate of absence due to illness. The question is how many of them are highly productive and how many are workplaces where the employees feeling of well-being is high. Perhaps a strategy of health promotion can create opportune conditions for the emergence of competence and involvement, without the workplace becoming one which "burns out" its employees through overstress.

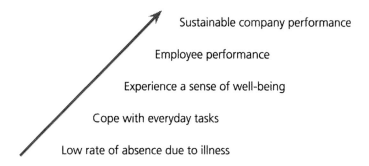

FIG. 8.1 *"Goal levels" in the organisation's health work*

Health promotion is a new concept and a new approach to thinking about health. It does not naturally belong to one particular discipline and thus it can function as a neutral platform which will allow new and different professional groups to play a part in health work. The development of health promotion work implies teamwork. Its tasks are made easier if no one in the team is, so to speak, "in their own domain" and has priority of interpretation when the health program is being designed. The personnel section, occupational health and safety department, senior management, trade union representatives and sometimes others as well can be involved: health work becomes essentially a project in co-operation.

Chapter 8

Is there some "best practice"?

When workplace health work has to be planned and purchased, it has been common practice that people from the organisation's personnel section have made an analysis of needs and discussed a suitable arrangement with the supplier. The supplier, in general, provides occupational health services, either internally as part of the organisation or as an external supplier. An agreement is drawn up with a list of services which can include everything from tests of human beings and environment to rehabilitation and educational services. On occasions, this purchase is carried out without any deeper analysis of the utility or value of the service being offered. The agreement has often been drawn up to conform with certain traditional notions about what such an agreement should include and what the employees involved expect.

With an increase in knowledge and cost awareness in the purchasing of health services, an analysis of needs becomes increasingly the basis for deciding what type of health work should be carried out. A growing number of workplaces supplement the analysis of needs with ideas and strategies for utilising the development opportunities which smoothly functioning health work can supply.

Irrespective of what questions are raised and how the analysis of needs is carried out, the workplace is often presented with a ready-made offer or program. It might be the company health department, together sometimes with the chief of personnel, who function as "experts" and are responsible for the content. Often it is the workplace's relevant senior manager responsible for financial and personnel matters, who has the job of seeing that the agreement is optimally applied.

This approach is efficient, since decisions, planning and purchase are carried out faster with fewer people involved. This form of centralised planning of health work is open to the criticism that in addition to the fact that it is oriented around problems and measures, it easily becomes ill adapted to the conditions of the specific workplace. Such a solution formulated by experts fails to give the individual employee a major role in participating and influencing how the health work should be carried out.

Experts know what influences health in the workplace and how one can prevent and cure ill health. There are also an increasing number of experts who have specific knowledge in the field of health promotion and who can propose initiatives which are good for health. When the need for-and interest in- the health questions of working life is at its height, more and more expert knowledge is published about e.g. the factors that determine health. There are books with concrete descriptions of working conditions, environmental factors, measures or behaviour which are beneficial for human health. The disadvantage with all these excellent pieces of advice is that they often remain no more than a "list of virtues" which enumerate things which, in general, are good for health, with a wide gap between the written word of the text and its application. In other words, they are correct and excellent descriptions. The problem is how do we translate them to apply to the unique circumstances of the specific workplace and how do we set about tackling things so that it turns out, as we want it to be. Seedhouse[142] holds that there is a lack of critical analysis of the role and value of educational initiatives in striving to bring about an improvement in human health.

The application of all knowledge and good advice about what is important for health in working life, or for health in general must occur via its interpretation and testing in the case of each unique set of circumstances. When someone writes a book in the field, it is usually a general description which applies -sometimes well, sometimes less well- to the specific workplace. The detailed

Chapter 8

description of cases or "Best practice" which is presented can have very good results when applied to the workplace for which they were designed. There is, however, no guarantee that the model is transferable to another workplace with similar results, at least not without careful reflection about what is similar and what is dissimilar in the conditions which apply to the two workplaces. Anyone, with experience of trying to bring about changes and development within an organisation, also knows that it is the presence of a capacity to deal with the local unique process, which decides if a change in the intended direction will come about or not. "Best Practice" can help with practical tips and have a motivating effect of a symbolic type, in the sense that the success achieved by others can serve to inspire everyone else with the desire to succeed. However, it requires knowledge of different kinds, the "right" preconditions and a wise strategy for bringing about the change in the direction we want.

For health promotion to become an established strategy for health in working life, its underlying idea and theoretical models must be well thought out and tested. Secondly, they have to be communicable and understandable, in other words, simple in structure. Thirdly, there has to be constant critical testing of its content and the consequences of its practical application. It is only thus that an original idea and a simple theory can be developed and be of value to society in the long run.

When we now suggest health promotion as an idea and strategy for salutogenic health work, the aim is not to give the whole world of health promotion a guiding model. Our starting point is that health promotion as a concept has existed in Sweden for several years. A major part of the interest in this form of health work is directed at working life and the workplace as setting, and it is on the basis of their particular circumstances that health promotion as a salutogenic health strategy has been formulated in this book and to some extent has also been tested. If this strategy continues to be successful and is allowed to develop, it can become a guiding model with both a normative starting point and a theoretical frame of reference. The experience, which is gathered, can then be applied to other settings such as the school and local society.

The contribution, which salutogenic health promotion can make, is not simply to appeal for a positive health-oriented approach with an emphasis on health factors. As we have mentioned earlier, what is important is to indicate the knowledge and ways of working which will help us to bring about a movement in the direction of improved health. In short, health promotion is what its Latin root suggests: *motion for (pro) health*. Seen in this way, the concept of health promotion is health work which is health- (rather than illness-) oriented and which pays due attention to the setting.

Health Promotion as three questions

Health promotion needs knowledge about health, about the conditions which determine health, and about how we are to bring about a change yielding better conditions for health. As a model to guide us, we can make use of three questions:

1. What is health?
2. What preconditions help to ensure that health develops and is preserved?
3. How do we bring about a change or development towards better conditions for health?

Health promotion as a practical activity, must deal with these three questions and dimensions of knowledge, which is something we can do today. We have already dealt with the first question in Chapter 3 and here we will not devote any more discussion to it.

Regarding the question about the conditions which contribute to health, there is a vast amount of knowledge about the working

Chapter 8

environment, working conditions, leadership and lifestyle which can help us to describe what we should strive after, in trying to create a health-promoting environment and a good workplace. Because the combined knowledge is so great and the need for adaptation to the specific set of circumstances is absolute, we can therefore benefit from having a model or theory like Antonovsky's theory of Sense of Coherence (SOC)[143] (cf. chapter 6). The model can help to guide us in understanding what it is, at the most general level, which contributes to human health. If we take this theory as our starting point for health work in the workplace, the model has to be translated so that it is clear what the three SOC dimensions - Comprehensibility, Manageability and Meaningfulness - mean in working life. It is only after that, that it is time to design the concrete content of the health promotion program. This can be experienced as a long-winded or circuitous process. But properly presented and after exercises in abstract thinking, it will be found that this theory is very stimulating both for the health worker and employees in the workplace. It provides greater understanding of the complexity of what it is that influences health, as well as an increased motivation for becoming involved in health issues. The depth and range of this kind of preliminary or preparatory work is determined by interest and the time available.

The third question - about how to bring about a movement for change- is the great challenge. How can health promotion be applied successfully in a workplace? How, given certain specific circumstances, do we bring about the process of change which will move us along the health axis to better conditions for health and therefore to better health itself?

Starting from the assumption that every organisation and workplace is unique and embodies its own particular preconditions and circumstances, it is wise, as we have said before, not to present some standard set program for the process of health promotion.

It would conflict with the whole idea of the importance of every employee's involvement and our emphasis on the actor-perspective, rather than the expert-perspective. There is, quite simply, no one single model which can be applied unconditionally and quite generally to all organisations and enterprises. We can, however, after several years of application in different places in the world, formulate a theoretical model with four criteria which describes what health promotion is or can be. It is sufficiently simple and easy to grasp to both indicate important presuppositions, and provide guidance when it comes to specific practical circumstances. This model is not a theory in the traditional meaning of the term, because it has not emerged from the empirical work of an individual researcher or group of researchers in a given project. The model with its four criteria is what has emerged from weighing up all the combined experience, presented in particular in WHO's Ottawa Charter[144] of 1986 and ENWHP's Luxembourg Declaration[145] of November 1997, which we have discussed in Chapter 4. On these occasions, both researchers and experienced practitioners have been involved in the arguments and although publications have appeared in connection with different conferences in different places, it is easy to note the central importance of the four criteria. These four criteria represent also many of the experiences which have been recorded and presented in various places in the world, during a growth in the work of health promotion in recent times.

Part of the aim of the present book is to make clear that health promotion is a field of knowledge and a way of bringing about change leading to better health, primarily in working life. As we know, there is nothing new under the sun and the approach to work which results from the four criteria which we will present in more detail, makes great use of the knowledge of the work on change which has been developed in working life since the 1960s. In this tradition, it is the emphasis on participation, process and pedagogy

Chapter 8

which is in closest agreement with some of health promotion's basic values.

In describing the four criteria, it is important to emphasise their role as a starting point and model for shaping the work of health promotion. The criteria serve as a guide, but they are far from being a standard "hands on" training manual.

Four criteria for health promotion

In the United Kingdom, the Health Education Authority[146] describes the essential character of Workplace Health Promotion as follows:

> ...a sustained program based on principles of empowerment and/or a community-oriented model using multiple methods, visibly supported by top management, and engaging the involvement of all levels of workers in an organisation is likely to produce the best results.

This quotation captures certain central characteristics. Health promotion means establishing a long term process which is clearly supported by the organisation's management and which allows each employee to influence the process. These are two important prerequisites in the processes of change which correspond very well with what Moldaschl and Brödner[147] call "reflexive intervention for sustainable change". Sustainable change requires reflection. It is advantageous to look upon health promotion as development or work aimed at bringing about change which is not a matter of applying "tailor-made" solutions. There is an idea, a basis of values and a number of criteria. The unique conditions involved in every set of particular circumstances, associated with an organisation or workplace, must then be taken into account when designing the process. It is the people in the workplace who – often with external

Health promotion in practice

support- must analyse, plan and carry out their own improvements for health. In this process, the four criteria provide a support.

We have earlier pointed out the four criteria in the Ottawa charter[148] and the Luxembourg declaration.[149] As a short recapitulation, we describe them briefly again.

Focus on measures promoting health

We work with an approach which basically emphasises health and its improvement. It is the notion of salutogenesis (to seek that which explains why health is preserved or improves) which is central and plays a great role in deciding which method of working and which forms of evaluation are to be used. This idea is followed necessarily by a wider view of the nature of health and what influences health. Health is more than merely not being ill. The determining factors of health must be sought both in human beings and in their surroundings. This allows a number of working methods and professions to enter health work.

Settings based

The *setting* is a place where something happens. In order to apply health promotion, we need to have a concrete situation or set of circumstances. No movement towards improved health can occur if we do not know and take into account the unique conditions and possibilities of the present situation. The determinants of health are not simply to be found in individuals themselves, but above all in their surrounding and particular circumstances. In the setting, in this case the workplace or organisation, there are the social interaction and the surrounding environmental factors which have to be analysed and perhaps modified in order to be more positive from a health standpoint. The workplace as setting has its own special conditions which distinguish it from places such as a

Chapter 8

school, a hospital, a housing area or any other category we choose for settings based health promotion. Every particular setting is a unique set of circumstances and smaller-scale workplaces must be treated differently from larger organisations.

One purpose of thinking in terms of setting is also to be able to single out this set of circumstances from the rest of a person's life. It becomes too complex to work effectively, if we have to take into account both the workplace and the rest of an employee's life outside the workplace. Thinking in terms of setting also entails "a system perspective" where analyses, decisions and activities are related to the individual, the workgroup or the organisation as a whole. This provides a structure which allows order and clarity. Another important goal which is prominent in the literature[150] is to have the work of health promotion incorporated in the ordinary activities and processes of the workplace or organisation.

Participation

Perhaps the most important success factor in health promotion, as indeed in all work dedicated to bring about change, is the individual's possibility of participating and exerting influence. In health promotion, the participation of all parties involved is both a value judgement that everyone's knowledge is relevant and important when decisions have to be made, but it also underlines that in the contemporary world, the participant affected plays an increasingly important role as a bearer of knowledge and a person who can exert influence. What is at stake is that decisions should be taken by those who are affected by them and this implies that power to make decisions, for example in an organisation, has been transferred from the top to those lower down. When we wish to bring about a change in the conditions affecting health in the workplace, success may well be decided by the extent to which we have succeeded in anchoring the proposal among the employees.

Health promotion in practice

If the process is dealt with wisely and everyone feels a participant, it is possible to create both motivation and energy which in turn contribute to making it successful. Conversely, a program which is run from the top, may lead simply to a shrug of the shoulders: "It's not our thing". Participation is based on skilful presentation and teaching methods– pedagogy– and on a balance between the fundamental structure and the creative process. . It is on occasions a difficult but necessary equation.

Process

The fourth criterion is the need for a process-oriented approach and describes how health work is to be carried out. It points to the fact that health promotion is more a movement with a certain direction, than the quick delivery of a tailor-made health program. In speaking of a process-oriented approach, we mean that there is an adaptation to- and influence on - the organisation's human processes, and yet, at the same time, some measure of control is not totally lacking, since there must be a structure and system to operate within. It is also based on continually checking up on what is happening, and adapting things so that attention is paid to the unique preconditions affecting the groups, the informal systems which affect the work, and the needs which appear later. Leading the process becomes an important function which is all the more demanding, the larger the workplace or organisation included in the work.

The idea of process reflects adaptation and flexibility in order to move in the best way possible from a particular starting point to a "desired state" or objective that has been set. In order to avoid ending up in unreflecting action with lots of "fun" activities, the three temporal dimensions of the change process – the past, the present and the future- must be taken into account.

Chapter 8

The four criteria of salutogenic health promotion can be seen as constituent parts of a conceptual whole, where each part is important. The parts must be combined with one another and integrated wisely to constitute a salutogenic health promotion strategy - a prerequisite for the movement towards a healthier workplace or organisation (Figure 8.2).

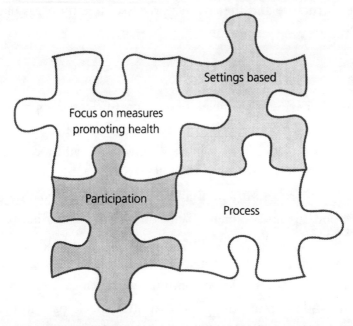

FIG. 8.2 *The four criteria of health promotion as constituent parts which are necessary in order to bring about a change towards improving the conditions which determine people's health in working life.*

What distinguishes salutogenic health promotion from promoting health in general?

What distinguishes salutogenic health promotion from promoting health in general and what is the relationship between them? We can say generally that the term 'health promoting' describes a property and effect. A product, an action, an activity, a measure or initiative, an environment and so on - all these things can be 'health promoting' or have an effect promoting health. Health can be promoted in different ways depending upon how the concept of health is defined and whose health is under discussion. For example, there is a difference between promoting an individual's health as opposed to a population's health. Health promotion can also be considered as the adoption of a particular viewpoint about how to look upon health work without specifying any conditions about how it is to be accomplished. Health promotion can involve a large number of measures of different kinds.

Without entering into a more profound philosophical discussion, we can state that salutogenic health promotion on the other hand is something more than the adoption of a viewpoint. Nor is it an effect or property in products or services. When we use our three basic questions as a starting point for salutogenic health promotion, we obtain a structure which both covers a more holistic approach and at the same time specifies and lays down guidelines for practical action.

1. What is health?
2. What prerequisites help to ensure that health develops and is preserved?
3. How do we bring about a trend towards better conditions for health?

It is assumed, even if this book has adopted a certain particular viewpoint and approach, that there are many different answers to all three questions. The questions, however, force us to a discussion which leads both to a certain normative position, the need for certain theoretical knowledge and a choice of practical approach.

Since salutogenic health promotion is based on a certain form of knowledge, we can see it as a field of knowledge, and eventually in the future, as an academic discipline. Salutogenic health promotion can also constitute a domain which according to Liss[151] contains ethical concerns, professional actions directed at a goal.

Chapter 8

In the practical situation to be found in working life, salutogenic health promotion can function as a strategy, a long-term overarching approach, for working with health in the workplace.

Summary

With its idea of salutogenesis and the four criteria, health promotion becomes a conceptual model for carrying out the work of change. This model does not differ essentially from work dedicated to change at a more general level. The point is that it focuses on health and the conditions of health and the fundamental values which follow from this, namely values which give priority to human beings and their opportunities for increased job satisfaction, well-being and health. Because of the focus on health, organisational and economic goals are of secondary importance. Nonetheless, since health is a precondition of performance, health work ultimately benefits also the enterprise. Conversely there have to be economic and organisational preconditions which allow the work of health promotion.

In this chapter, we have described four criteria which collectively form the basis of health promotion work.

- *Focus on measures promoting health* – A salutogenic approach which also forms a normative basis in health work.
- *Thinking in terms of setting* - The concrete situation and its preconditions for operating with a health promotion strategy. System thinking and integration is then employed.
- *Participation* - Human conditions which must be satisfied.
- *Process* - the form for the work of change

Health promotion in practice

With these criteria, we have a framework for operating within. They contain each and every one of them, many aspects and explanations which the professional health worker needs to be familiar with and to use. What are the implications of these criteria and how do we apply them in different situations? Is there, for example, a least common denominator which we must bear in mind when we are deciding about the conditions and level of participation in a certain workplace? In the individual workplace, our own knowledge and understanding requires to be supplemented.

In the next four chapters, we shall describe things in more depth and hopefully both clarify and draw attention to certain finer points in the case of the criteria, which can make it easier to apply them.

9 Focus on measures promoting health

It can appear self-evident that health promotion should aim at wellness. We have already described health promotion and the idea of salutogenesis on the basis of Antonovsky's books and the international tradition.

What needs to be added is a description of how we can regard health promotion in its practical application, when the idea of salutogenesis is to be introduced in a situation where different forms of health work and different established professions and traditions have to co-operate. As a result, this criterion requires some commentary and clarification which facilitates dialogue and adaptation to practice in different situations.

In a report in 2002[152], WHO states the following:

> For the successful development of workplace health promotion, it is important to identify factors contributing to development of health and to facilitate and strengthen impact of such factors conducive to the health of all staff (p.27)

This text describes the salutogenic work and coincides with what we mean by salutogenic health promotion and constitutes the most important demarcation of the field of health promotion in relation to other health strategies.

When we focus on positive measures encouraging health as the criterion of health promotion, as opposed to curative or preventive measures against ill-health, we obtain a demarcation of the field of

Chapter 9

health promotion as one which has a salutogenic goal i.e. one geared to health and wellness. The question of the origin of health or of what helps to ensure that certain people, despite stress and vulnerability nevertheless preserve their health, becomes the critical identifier which most clearly distinguishes salutogenic health promotion from other forms of health work. The approach of seeking these health factors, or resistance resources, is something other than seeking the causes of illnesses or explanations of why certain people become ill. It is a quite different way of posing the question and it entails a quite different way of looking at health. Consequently salutogenic health promotion must also be a different way of conducting health work. This delimitation and clarification of the subject is necessary for several reasons. If health promotion and other health work focussing on health rather than illness are to be able to develop, certain limits must be imposed upon its range of knowledge and activities. For the health promotion professionals, as it is now or will become, this is an issue of identity: who are we? What can we do? What is it that we deal with? Other health professions also need their concepts to be put in order either to extend their activities or to work together with the new "health promoters". When health work takes place within an organisational framework, interdisciplinary co-operation is a "must". The issue of health concerns not only the traditional occupational health professionals and the relatively new health educators, but also trade union representatives, the occupational safety organisation, and the personnel department, management and in certain cases the organisation's board or governing body.

People advocating the idea of salutogenesis and the merits of health promotion, must be constantly aware that the salutogenic and pathogenic perspectives complement one another. Both are needed and together they help to ensure that we can work for health on the basis of the three strategies: cure, prevent and promote. Curing or alleviating illnesses or injuries as well as solving emergency problems

Focus on measures promoting health

can be a necessary preliminary step before health promotion work can come into play.

From an initial perspective based on illness, health work builds up knowledge about what causes illness (pathogenesis). This knowledge in turn generates two strategies for dealing with the problem of illness namely, the treatment of the problem that has arisen and preventive measures to eliminate the risk of illness in the future.

The first of these two strategies is focused on alleviating, treating and curing illness. It is pathology, that is to say the study of illness, which determines the choice of measures and the health worker has to be trained in this area to be able to choose the appropriate medicine and the appropriate therapy.

It is somewhat more difficult to draw boundaries between the concepts of illness prevention and health promotion. This is often the point where the discussion and understanding of the specific characteristics of health promotion becomes stuck in a groove. Is it not obvious that a health promotion initiative is also necessarily preventive? If the work of health promotion is successful, does it not inevitably mean that fewer people become ill? Is that not exactly the same case throughout? It has to be admitted that sometimes these concepts seem to coincide.

In order to make clear if a certain form of health work is to be classified as preventive or health promoting, one way is to ask why it is being carried out. The motive for a health initiative can, for example, be that we have noted an increase in the occurrence of hip fractures in older women. Irrespective of whether we then choose to prescribe calcium, oestrogen, or increased physical exercise, our starting point is the knowledge about the risk of hip fracture among the elderly and the measure we adopt is one of prevention. At the population level, on average, it is possible to note the value

Chapter 9

of a preventive measure, but in the case of a specific individual, the initiative was perhaps completely unnecessary, since the person might have avoided illness in any case.

When an initiative is carried out to strengthen people's ability to cope with life's stresses in general, we call it a salutogenic initiative. It is a less precise approach. We do not know beforehand what value the strengthening initiative will have; we only know from experience that it can be a resource in various circumstances. It is here we find a dividing line between prevention and promotion. Prevention is in general more precise in its aim; it is based on knowledge and the belief that, exactly as in the case of curing an illness, it is possible to find a measure or remedy which is fairly exact in its effect. In his first book[153], Antonovsky called this "the Magic Bullet Approach". At the same time, he criticised medicine's belief in its own capacity and held that medicine's ability to "hit the target" is not at all as great as we would wish. According to Antonovsky, we can also hit the target by accumulating knowledge about what helps to preserve and improve health and then to pursue promotion work to support and strengthen these *General Resistance Resources* (GRR).

Probably the majority of people working with health, are keen that their efforts to help the state of health, both of the individual and of the population at large, are as good as possible. This is based on the fact that we look upon health as an important asset for all human beings and one to which we attach high priority. As a result, it is meaningful to participate in such work. It is considered that the strongest driving force in people's work is to be allowed to contribute to creating those things which are valued most by other people.[154] The majority of health workers - if they were to be asked the question - would also certainly reply that they adopt a health promotion attitude and approach which at a purely conceptual level, may very well be the case. Both the concept of health and of promoting health have, as appears from what we have said above, different meanings for different people and for different professional groups. There is no accepted scientifically or theoretically applicable definition, but rather as Brylde and Tengland[155] put it, it is a matter "of finding a definition of the term which is appropriate relative to certain goals". (p. 170). If there is a reason for-or interest in- trying to clarify what is to be called health promotion or not, the discussion can be taken up in every specific setting. Is it health promotion from a salutogenic point of view or is it merely a positive approach to, or general interest in the health of the population and in improving the epidemiological statistics? The positive health approach, that we wish to call health promotion, implies that the improvement of health is based on finding the factors which promote health in human beings and their living environment and strengthening them. It is this that defines the salutogenic approach to health promotion even if the person or group the health worker meets is defined on the basis of ill-health (pathogenesis).

Can language and the use of concepts have any practical significance? Linell[156] holds that every professional group has their own professional language which underpins its professional identity and the attitudes reflected in its concepts and values. When different professions have to co-operate, conceptual obscurities or the lack of empirically viable concepts can create misunderstanding and conflicts with other professions which are active in the same sphere.

The word "patient" is an example of a word which reflects a viewpoint where people's own resources are not traditionally treated as a primary contribution to better health. Being a patient is a role where the setting and how one understands it, has great importance for how people regard their role and their own actions. The setting and role steer their thoughts and presumably influence both the experience of the health work which they form part of, but also the extent to which they themselves can and will be active in improving

Chapter 9

their health. In English, the very word *patient* reveals perhaps something important about this role.

In the practical work, it is therefore necessary to devote time to dialogue between the representatives of the various disciplines so that the different perspectives and disciplines can arrive at some understanding of each other and strive to find an agreed use of terminology which supports both the idea behind the work and its content.

There will be more than one meaning of concepts such as that of promoting health when different professions work together in the same area. The word *health promoting* is, in its logical meaning, a salutogenic term. However there are many who, despite the fact that they adopt a pathogenic perspective in their health work, would hold that they are carrying out health-promoting measures. This is illogical in curative work, as long as the goal is to cure and the goal to be evaluated is to be free from illness. However, it can be somewhat more logical at an overarching level when we note that all health work aims at improving the population's health in general. The only problem which remains is that even this goal is still evaluated using medical (epidemiological) measures. If the measures were to deal with functional capacity or quality of life, the health promotional logic would be clearer. There is every reason to aim at the goal of isolating and developing a salutogenic strategy as a complement to the pathogenic strategies which have been in existence for a long time.

The holistic viewpoint

The salutogenic viewpoint entails an enlarged view of health and health work. This holistic view can be described in several ways. We can note that we have both pathogenic and a salutogenic perspectives and in health promotion we suggest to focus on the latter, but we must look upon both as necessary and complementary. We also have a holistic view of human beings which means that we take account of both body and soul. We have earlier described health in dimensional terms of "experienced" health and "measurable" health. The physical body and mental consciousness influence - and co-operate in - the human organism. Health promotion sets out to bring together human beings and their situation or setting where environmental factors, social circumstances, mental processes and physiological states interact and influence the experience of well-being as well as the occurrence of stress reactions.

The holistic approach entails that the fields of knowledge encompassed by health promotion which we have previously described, range over a wide area and several disciplines. Health promotion is necessarily a field of activity which covers several professions. When we take the workplace as its setting, knowledge is required about the whole system from human physiology to how work groups function and how the working environment and organisation are to be dealt with. There are individuals with a good education and long experience of the work of health promotion, but in practice, health promotion is not a one-man show. The holistic standpoint presupposes that several actors participate in the process: it is a multi-professional and interdisciplinary field.

Holism has its counterpart in atomism, which implies that one specialises in the small constituent parts. This approach is also necessary both in developing knowledge and as a *modus operandi*. What is important is not that there is an opposite or counterpart: it is more a matter of applying the right strategy depending on the nature of the task in hand and how we define the problem. The two approaches ought to be complementary if we look at the broader picture of health work.

Chapter 9

Focussing on the whole does not remove the necessity of also being conscious about the constituent parts. It is, however, a choice of starting point when we say that health promotion proceeds from the whole or from the specific setting. In health work where the great part of the staff is made up of generalists rather than specialists, it is the way of working practically and of developing knowledge, which serves as a guide. The whole, the wider spectrum and the particular circumstances, are also closest to the most important actor in health promotion, namely the individual person.

In the Luxembourg Declaration[157] of 1997 there is an introductory passage which points out the importance of holistic thinking in health work in working life:

> Workplace Health Promotion (WHP) is the combined efforts of employers, employees and society to improve the health and well-being of people at work.
>
> This can be achieved through a combination of:
>
> - improving the work organisation and the working environment
> - promoting active participation
> - encouraging personal development

Health in the workplace presupposes not only a combined effort from society, employers and employees. We also require to create preconditions for health and well-being through an improvement of the work organisation, working environment and by supporting people's personal development. It is a question of an attempt to tackle things as a whole, which involves much more than merely answering to legal requirements relating to rehabilitation and the working environment.

Focus on measures promoting health

Official documents from the European Network for Workplace Health Promotion (ENWHP) describes Workplace Health Promotion as a "corporate strategy" which involves both preventive and health promotion initiatives. WHO notes that this is the case in its document *Good Practice in Occupational Health*[158] which cites and summarises the Luxembourg Declaration:

> Workplace health promotion is seen in the Luxembourg Declaration as a modern corporate strategy that aims at preventing ill health at work (including work-related diseases, accidents, injuries, occupational diseases, and stress) and enhancing health promoting potential and wellbeing in the workforce. Expected benefits for workplace health programs include decreased absenteeism, reduced cardio-vascular risk, reduced health care claims, decreased turnover, decreased musculo-skeletal injuries, increased productivity, increased organisational effectiveness and the potential of a return on investment. However, these improvements do not have to be long lasting, and require continuous involvement of employees, employers and society. (p. 28)

Internationally Workplace Health Promotion is predominantly described as containing both preventive (pathogenic) and promotive (salutogenic) forms of health work. This book takes a different line by advocating a concentration on a salutogenic promotive strategy. In other words, it does not rest upon pathogenic and preventive ideas. By thus concentrating on health promotion as a salutogenic strategy, we define more clearly where the subject begins and ends conceptually and empirically and thus create better preconditions for developing knowledge and methods for health promotion work.

This approach is supported, for example, by Ilona Kickbusch[159] who already had a central role in WHO at the time of the Ottawa

Chapter 9

Conference and who has since continued to develop and stress the need for a focus on promotion:

> In my view, health promotion is "determinants based". By this I mean to express that it bases its strategies on best knowledge of how health is created and how social and behavioural change is best effected. It aims to maintain health as a resource and prioritise investment in health through the following four questions.
>
> What creates health?
>
> Which investment creates the largest health gain?
>
> How does this investment help reduce health inequities and ensure human rights?
>
> How does this investment contribute to overall human development?   (p. 267)

Polanyi et al.[160] follow the same line in the concept of Promoting Workplace Determinants (PWD). They describe an approach which is akin to the salutogenic health promotion, in the following way:

> ....these approaches involve a structured yet democratic process of issue identification, analysis and action. In the workplace, the successful implementation of such a process would ideally include, to the greatest extent possible, a/ broad-based commitment and participation of both workers and management in all stages of the project, b/ an openness of worksite participants to deal with the full of internal and external determinants of health, c/ targeting health issues that are a priority of workers (and likely significant epidemiological), researches acting as technical resources and process facilitators, e/ both quantitative and qualitative research methods, f/ a long term commitment, and g/ ongoing evaluation of both process and outcomes to feed back into project planning. (p. 147)

The salutogenic strategy for health work in working life may continue to define its prerequisites and conditions which may then be analysed in the light of other health work. It is hard to know how far there will be a development so that everything that is now in general terms "health promotion", will in the future be salutogenic in character. Perhaps it is neither desirable nor necessary, provided the various actors and professions can work together and see that different types of initiative and competence are needed in health work. It isn't a question of something being categorically right or wrong: rather difference and multiplicity are assets in a creative process of development, as long as there is critical evaluation. All the time there has to be a conscious analysis of new "truths" and new concepts of health. What values and which policy steer things? This question applies both to the practitioners working in the field and to those who determine research grants and how resources are divided among the various activities.

A case of tunnel vision?

Medical health work has historically been problem-oriented. It is only when illness has manifested itself that the health worker has had something to deal with. The problems have mainly appeared in the human body, at least if we look upon mental problems as a preliminary stage of somatic illness. Belin holds that in our technocratic society, we are particularly prone to interpret everyday irritations and human problems of every kind, in somatic and medical terms.[161] Medicine has also gratefully received these somatic problems and created therapies and pharmaceuticals in order to treat the part of the body which exhibits some symptom or dysfunction.

Chapter 9

The problem oriented way of dealing with things is also predominant in health work in working life and is illustrated for example in the Luxembourg Declaration[162]:

> All measures and programs have to be oriented to a problem-solving cycle: needs analysis, setting priorities, planning, implementation, continuous control and evaluation (project management). (p2)

This description may testify to a well structured and thoroughly considered way of working. It may also testify to a form of tunnel vision in always taking the problem as the starting point for our actions. Mari Kira[163] cites Ludema et al. who direct their fire at this traditional Western way of carrying out development work in organisations and societies:

> If we devote our attention to what is wrong with organisations and communities, we lose the ability to see and understand what gives life to organisations and discover ways to sustain and enhance that life-giving potential (p. 112)

If pathogenesis and problem-solving have hitherto represented one form of tunnel vision, there are other examples of the phenomenon elsewhere. There are dedicated "health missionaries" who espouse the salutogenic approach and who believe that they have the best- even the one and only- solution to how human beings can have better health. "Let us simply introduce" they say, "positive thinking along with a development model geared to possibilities and we shall obtain outstanding results". This kind of tunnel vision may depend on the ideological undertone which sometimes has accompanied health promotion. Concepts such as holistic thinking, positive thinking and the idea of development are, according to

Focus on measures promoting health

Korp[164], part of a ideological formation which is natural when a professional or interest group wants to create some kind of group identity, and hence also, to emphasise what is specific to them as a group as opposed to other competing groups.

There are different ideas about how we can work with health. These ideas generate strategies as a means to application. Knowledge and methods of working with health - curative, preventive and promotive- are being developed all the time. A possible brake on development can arise when different views and different meanings attached to the concepts used occurs in the respective idea and strategy. Seedhouse[165] maintains that the development of health promotion has also been impeded by lack of theory, something which he himself tries to repair. There is reason to reflect more about whether health promotion is being developed in a wise way or whether it is impeded by an absence of theory and conceptual confusion. This reflection is a challenge to researchers and practitioners to invest more time on developing theories and critically reviewing what we are engaged in.

A continued dialogue and critical analysis of everything going under the heading of health promotion can hopefully lead to an increased degree of mutual understanding. The medical curative and preventive approaches have had several hundred years in which to work out their respective ideas and to fashion their conceptual systems. It may well take a little more time for the health promotion approach to find its feet as regards forms and concepts. After all its history dates back only fifty years.

If this book's proposal to limit health promotion to health promotion based on the salutogenic approach were to be accepted, it would mean that the history of the subject begins in earnest with Antonovsky's book[166] in 1979. This was further emphasised by Antonovsky in an article in 1996[167] where he proposed the

Chapter 9

"salutogenic model" as a theory which could serve as a guide for Health Promotion.

Focus on promotion fits in with the development tradition

Within the organisational and corporate world, there is a long tradition of work dedicated to development and improvement. Within production and commerce, the whole of the modern period of industrialisation and modernisation has been characterised by a tradition of improving efficiency, productivity, quality and so forth. The history of organisations contains many ideas and schools of thought which at various times have proposed new solutions, mainly aimed to benefit the companies. Despite certain epochs being bracketed with a movement for more humane rationalisation[168], it has seldom been the case that changes have been introduced primarily for the well-being of the working man or woman.

There is not space in this book to go into this vast area in greater detail. We merely note that in this tradition there are theoretical models which can serve as objects of comparison for the process work of health promotion. Often there are no great differences, methodologically speaking, between developing an approach aiming at superior quality production and developing conditions to secure good health. In both cases, it is a question of working constantly to improve things by creating and optimising conditions. Both the quality of the product and the employee's health are, pre-eminently, multifactorial phenomena.

Choice of strategy and work model is a question for both the quality and health representatives. Is it really smart to analyse the problems and weaknesses which exist, to deal with them and in the future to prevent them? Would, perhaps, not a better alternative be to investigate which factors have the greatest potential for providing better quality or better health, and devote ourselves to developing them and seeing that they are preserved?. The quality representative and probably also the health representative will finally conclude that it is wise to use both strategies. How exactly is the situation in this particular workplace? We choose the strategy and work approach which we believe will be successful on the basis of local conditions and a collection of unique circumstances.

Finally there is always a tradition or habit or "local chauvinism" which may prevent a choice of strategy which is appropriate to the purpose in hand. One encounters the attitude that "no one is going to come here and teach me. I know all about this and I am going to do it the way I always have". This type of attitude may be dealt with in the same way as in all other work involving bringing about change.

What is to be promoted?

There are never sufficient resources to allow us to deal with all aspects which are important for health. We perhaps do not even know what is really good for health. Whose health is involved? However, the question remains and needs to be answered: what is to be promoted? What factors have the greatest importance in influencing human health positively? In some way, the agenda of the health promotive process has to be agreed on. We need to know what means are available, in order to achieve our goal.

It is a question of where this knowledge is to be found or who has it. Both the goal set and the program devised to achieve it need to deal with political/moral questions about what open or hidden aims influence health work and what positive and negative consequences a health program can entail. In order to deal at a practical level with this selection of health's determining factors, there are certain starting points. An initial question which is discussed by Seedhouse[169] is the extent to which the health promotion process is based upon what has been scientifically shown to be good for health (evidence based) or if it is other values or interests (value based) which determine the agenda. Seedhouse holds that health promotion is, irrespective of setting, basically political and it is important to accept that, to be aware of it and then to design the work of health promotion on the basis of clearly thought out values and wellfounded experience.

When, in the case of a specific concrete setting, the health promotion process is to be designed, there are at least three ways of producing a list of what is to be promoted:

- with the help of theory
- with the help of experts
- with the help of those who are affected by the health measures

Health promoting measures- with the help of theory

As set out in Chapter 6, Antonovsky's theory of SOC is a theoretical guide to the conditions which can help to promote human health. It is the product of empirical investigation and can therefore be said to be based on practice and thus also can be used in practice.

It is a more comprehensive procedure to begin with theory and on this basis determine the factors and activities which are to be selected. For the research worker, such work is natural and often necessary. It ought to be of interest and rewarding for the health worker to illuminate and criticise their practices with the help of a theory. As a practitioner, the health worker is also in the position to test if the theory is relevant.

SOC is also a rewarding phenomenon which is worth studying and discussing by everyone. In this theory or model, scientific knowledge and the experience of people at large are brought together, making the drawing up of the agenda a stimulating and participatory process.

Health promoting measures- with the help of experts

Perhaps the most common way of doing things when a health program has to be designed, is to ask the experts in this domain. The experts can be professional actors e.g. company health and safety personnel, organisational consultants or a national health institute which provide information about what is good for human health. Expert knowledge is an excellent thing to have access to, but if isolated from the everyday world, expert pronouncements can sometimes be quite foreign and inappropriate to the task in hand. Research is certainly right to claim that less fat food, a veto on tobacco and regular physical training are good for health. There is, however, no proof that such a life style is the primary choice for all people in all situations.

A usable form of expert knowledge is that which is capable of initiating and dealing with the process oriented approach which is a criterion in health promotion. This is more about "knowing how" than "knowing what", but process management also involves dealing with and integrating scientific knowledge, expert advice and common sense. The role of process manager, therefore, requires a breadth of knowledge in the health domain to allow communication with all the actors participating in the process.

Chapter 9

Health promoting measures - with the help of those affected

Ask men and women themselves - the people it concerns - what they think the health promotion program should involve. In many countries, general awareness and knowledge about the determining factors of health is quite extensive. A creative SWOT -analysis (strengths, weaknesses, opportunities and threats) or a survey questionnaire can serve as a preliminary activity in order to describe the current situation and to choose collectively those factors which are to be promoted.

Today, it is an accepted axiom that the people who are affected must not simply be talked about in the third person: they need to be addressed and asked for their opinions in the matter. When people themselves are involved in drawing up the agenda, the finer issues of participation, acceptance and motivation, are more likely to be resolved, while simultaneously the health program itself is better adapted to the situation in hand.

There is no need, however, to refrain from availing ourselves of expert knowledge and tested experience which may be needed to import knowledge and establish priorities which ensure that the health initiative is kept on course. Individuals' hobby-horses can risk taking up too much attention and transform a health promotion initiative aimed at change into a forum for sectional interests.

Choice of level for initiatives

It is impossible to get away from the well established experience that the individual's interest and personal responsibility form necessary ingredients in a health promotion program. One reason is that it is individuals themselves who experience if the health promotion work has been a success or not. It is also well established that an initiative which is focussed on the individual is more or

Focus on measures promoting health

less an obligatory component in order to create the preconditions for health at the individual level. In other words, the experience of health is located at the individual level. It is at this level that one finds perhaps the most effective possibility of influencing both the preconditions for health and the experience of health in itself. Thirdly, it is at the individual level that one finds the interest and willingness to become involved in a workplace that promotes health. If the employees themselves are unwilling to become involved, then no one else can carry out the work of health promotion in a workplace.

In order to create good conditions for health, the health promotion work must go beyond the individuals themselves and also deal with the group and organisational levels. According to Rimann and Udris[170] it is the structural preconditions and the functional life setting which embody many of the conditions influencing health. For that reason, in the health promotion process in the workplace, there has to be someone who ensures that the initiatives and responsibility are distributed among the various levels of the organisation.

What is;
- the responsibility of the company/top management?
- the responsibility of managers?
- the collective responsibility in the workgroup?
- the responsibility of the individual?

The person leading such a discussion needs to ensure that the aims and means of the health work in question are critically analysed, that the input of basic knowledge which determines the choice of where to target the initiative, is wide and rests on science and tested experience, while at the same ensuring that attention is paid to the individual person's possibility of exerting influence.

Chapter 9

The consequences of promotion

How, in spite of everything, do people still quite enjoy their workplace? The question sounds like the original salutogenic question: how is it that people, despite strain and stress, still retain their health?

What are the consequences of asking questions in different ways? Can it lead our brains to start thinking along new lines and make people look at their surroundings and existence in a completely different way?

There is a tradition where most work dedicated to improvement is based on problem analysis and the accompanying program of measures, human attitudes perhaps become in general more concentrated on problems and problem-solving. Do we see the glass as half empty or half full? Why is there an acorn lying on the ground? Is it because it fell down from the tree or is it because it will begin to grow, set its roots in the soil and mature into a new oak?

Some people are more concerned with possibilities and are ready to make use of and promote what is positive in life and which flourishes, despite all the adversities. Is that something that we simply decide on our own? Or is it an attitude which is just as much part of the culture of society and organisations, as it is a part of the individual human brain?

This can sound like some topic for a lecture by a consultant preaching the gospel of optimism. But there is still reason to ask the question, since people in different countries and cultures look differently upon existence and about how life should be lived. Living with the focus on salutogenic health determinants can sound grandiose, but is nevertheless what it is ultimately about. Besides the major problems and development projects, there are masses of minor factors which at first sight seem banal or unimportant. Despite the fact that they are minor, they contain great development potential, just like the acorn. It is a matter of being receptive to the picture as a whole and being able and desiring to see both great and small health factors as something that it is worthwhile investing development money in.

A focus on promotion is also a focus on possibilities and very often such an attitude tends to have a positive effect on job satisfaction, well-being and health.

Summary

Focus on promotion is a direction and a choice of the way ahead. Just as there is a focus, so there is a background which also has to be borne in mind. The idea of salutogenesis is central to health promotion: it is the health factors which are the most important. At the same time, health work must be adapted to the unique demands and conditions of each situation or setting. For that reason, although the promotion approach is the main focus, it is carried on in conjunction with the preventive and curative approaches.

The focus on promotion presupposes a holistic perspective with regard to human beings and their situation. On the basis of the picture as a whole and the particular situation, we shall proceed to extend our ideas with the help of the other three critical components of health promotion - namely setting, participation and process. In the next chapter, we shall examine the concept of setting.

10 The workplace as setting

Health promotion work deals both with human beings themselves and with the conditions governing their lives. For that reason, the issues must be tackled by taking into account the situation as a whole. WHO discussed this principle in the Ottawa Charter of 1986[171]. *"The inextricable links between people and their environment constitutes the basis for a sociological approach to health."* (p.6).

In practice, it is impossible to cover at one fell swoop human beings' whole environment in the work of health promotion. For that reason, thinking in terms of settings has been of help in delimiting and describing the conditions for the situation we are working in. What we call "setting" is the organisational unit or geographical place which constitutes the underlying context in which the health promotion work takes place.

In the WHO Health promotion Glossary[172], the term *setting [for health]* is explained as follows:

> The place or social context in which people engage in daily activities in which environmental, organisational and personal factors interact to affect health and wellbeing.
>
> A setting is also where people actively use and shape the environment and thus create or solve problems relating to health. Settings can normally be identified as having physical boundaries, a range of people with defined roles, and an organisational structure. (s19).

Chapter 10

On WHO's initiative, health promotion is applied within social settings such as school, workplace, hospital, prison and housing areas and has the task of creating within them what we call "supportive environments"[173].

From focus on the individual to focus on the context

The Ottawa charter resulted in a stronger focus on the development of supportive environments for the benefit of health within health promotion. The work of public health was supplemented by "community development" where health promotion covers not only health education and human life style, but also extends its perspective to include human beings and the social, cultural, political and economic circumstances of their lives. This could be seen as a new idea in health promotion, but it was in fact a traditional and well established way of thinking about things in disciplines such as social psychology and sociology. Wenzel[174] maintains that thinking in terms of settings should have meant that the social sciences had a greater role to play in Health Promotion than was subsequently the case. There ought to have been a clearer re-orientation from the health problems of the individual and thinking about risk factors, to more thinking in terms of systems and treating organisations as the complex phenomena they are.

Despite thinking in terms of settings, a great deal of health promotion would continue to be concerned with individual lifestyle and behavioural change. Wenzel cites Baric who criticises thinking in terms of settings, holding that:

> ...settings are defined frames of reference for any type of intervention as regards health promotion/ health education. People are seen as objects of intervention programs, they are targeted to persuade them of behaviour changes, and they are

defined as bearers of statistical data rather than as individuals and collectives with specific traditions, biographies, needs, experiences, and patterns of behaviour having been developed over the course of time. (p.2)

Baric makes an important point here. How do we regard the person or employee in the setting in which we are pursuing our work of health promotion? Is thinking in terms of settings only a way of delimiting the living world of a human being which makes it easier to specify goals and treat employees in terms of statistically based arguments about lifestyle? In this case, health work is still a pure lifestyle issue in health education[175] where the aim is to persuade the individual to adopt a healthier style of behaviour. The individual bears the blame for the risks of ill-health and is responsible for altering his/her behaviour to a sounder one. The expression "blame the victim" is usually used to capture this viewpoint.

Thus decision-makers and health workers should have the obligation to extend their perspective by moving from the responsibility and conditions of the individual to cover in addition the social context and the organisational conditions. With such an extended view, thinking of settings becomes a framework and basic structure which, at several levels, provides a prerequisite for *health promotion*, a strategy for bringing about a movement towards improved conditions for health. When the health promotion work is targeted at a setting and thus simultaneously involving several system levels -the individual level, the group level and the organisational level - the preconditions are thereby created for a better result than if each system level was dealt with on its own. According to Poland, Green and Rootman[176], health promotion can help people to work collectively to alter circumstances which lie outside the particular control of individuals, groups or company management. It is co-

Chapter 10

operation and integration between the organisational levels which provides the best preconditions for a change.

The setting is not primarily a simply defined and delimited situation or a geographical/spatial surface, but is rather a complex system where several parallel processes and organisational levels are mutually dependent and influence one another. The people within the setting form part of the context and must therefore participate in the work of change. When it is at its best, thinking in terms of setting implies that individuals, instead of being the objects of an initiative , acquire the role of actors, where their knowledge and involvement form the most important contribution in the development of good health conditions. People within the setting, who act collectively, have the best possibilities of influencing the surrounding environment. The transformational work of health promotion, when it is successful, can change a workplace both physically and socially, to become a better place to be and one in which it is easier to retain one's health. Health promotion, by thinking in terms of settings, is at least more effective than single isolated activities aimed at improving health, if we wish to achieve a long term improvement in the health conditions in e.g. a workplace.

While individuals and groups can influence and change their workplaces, their behaviour is partly a result of the environment they find themselves in. The culture, tradition, activity etc of the workplace shapes or limits the behaviour of individuals and groups. The workplace is an obvious example of a setting where the context both steers people and is steered by them in continuous interaction. This interaction creates the conditions for health which the work of health promotion is aimed at. Within the current framework we prepare a change to a new, better state. When we attain it, we have created a new framework and conditions, from which to take the next step towards improvement.

The workplace as setting

A setting with several interested parties

The importance of the workplace as a setting for health promotion was emphasised at the WHO Jakarta Conference in July 1967, in a declaration which proclaims among other things[177]:

The participants attending the Symposium on Healthy Workplaces at the 4th International Conference on Health Promotion (Jakarta, July 1997) underlined the great importance of work settings for the promotion of health of working populations ...Comprehensive workplace approaches are essential which take into consideration physical, emotional, psychosocial, organisational and economic factors both within work settings and all other settings, in which people fulfill their multiple life roles... This approach is based upon the following four complementary principles: 1. health promotion, 2. occupational health and safety, 3. human resource management, and 4. sustainable (social and environmental) development.

These latter four points can be seen both as principles or functions, and as a listing of interested parties and actors who can or ought to participate in the health work. It is important to clarify the actor issue in particular when we are working with the workplace as setting. The larger the workplace, the more interested parties there are who can, in different degrees, be actors and influence both the conditions governing health and the process towards a healthy and sustainable workplace. In this context, the health work actors greatly exceed the numbers of those who are traditionally classified as health workers. Depending among other things on the size of the organisation, the health work is potentially linked to a selection of the following:

Internal interested parties

• The organisation's board and corporate management

Chapter 10

- managers
- human resource department
- occupational health and safety service
- trade union organisations
- work groups
- employees

External interested parties
- trade union organisations
- employers' organisations
- public authorities (legislation/ regulations)
- local society (local and municipal)
- Society at large (national actors and EU)

Customers, suppliers, partners, clients, patients or others who in one way or another influence, or are linked to, the activities of the workplace, may also in different degrees be reckoned among the interested parties.

Co-operation

Health and productivity in work are linked. A sound workplace with healthy employees performs better. Health can therefore be seen as a strategic resource in the organisation. The task of creating optimal conditions for health is thus a strategic issue, or at least ought to be one.

Awareness of the importance of health and the role of the organisation in health work varies greatly between different workplaces. In most countries, the employer's legal responsibility for the work environment has led to greater involvement, above all,

The workplace as setting

in medium and larger organisations. In many places, the concept of systematic work on workplace environment has entailed a more conscious treatment of physical and psychosocial issues linked to the working environment. More and more organisations choose also to apply a more promotion-oriented approach and there are good examples of organisations where health work has increasingly become a strategic and integrated issue.

In many workplaces, however, there is neither awareness nor activities in the sphere of workplace health and environment. The situation is worst in smaller workplaces where there is a great need for improvement. At the ENWHP[178] conference in Lisbon in 2001, special emphasis was placed on the importance of health work in small and medium sized companies.

Irrespective of where they stand in this comparison, all workplaces and organisations have reason to reflect over their views regarding the following questions:

1. Has the health of the workforce importance for our performance?
2. In this case, is the creation of optimal conditions for health an issue of strategic importance?
3. Who is formally responsible for the health question in our organisation?
4. Who is actively pursuing the issue in our organisation?

When the question of the conditions for health is of strategic importance, it ought also be decided and pursued at a strategic level. The organisation's strategic level has also the power to determine the agenda and the allocation of resources. An aspect of importance is also that whoever has power, can also decide how much power and influence should be delegated to other interested parties.

Chapter 10

In working life, there is tradition of dividing people into parties and negotiating about resources, influence and co-operation. Many issues are decided in discussions between the two main parties; the employers and the employees where the latter are represented by their trade union organisation. Health promotion is based on a "win-win" approach: both the employers and employees win by creating good conditions for health in the workplace.

In the work of health promotion, a group made up of the parties, can be a suitable forum for discussions of health work. However, the aim should be to involve more people than the traditional *parties* by reflecting about the various *interest groups* involved with the organisation and able to contribute in some way. The organisation which manages, at an early stage, to establish an honest and open co-operation between the various interest groups, is best placed to succeed in achieving a good result. The determination to contribute to the work is ultimately influenced by how far people are convinced that the influence and value resulting from their input, is distributed in a just manner.

When inevitable conflicts of interest are tackled wisely and the collective value is made clear, there is in general an excellent basis for co-operation.

Health promotion presupposes co-operation and teamwork throughout the whole process from the initial preparatory work, analysis and planning to the implementation of the measures. It is the various interest groups in the organisation and the various professions with their special competence with respect to human beings, organisation, health and change, who have to co-operate.

Every workplace is unique

Perhaps the most important principle in health promotion is that the health of human beings, their well-being and job

The workplace as setting

satisfaction does not only depend on their own actions. Their life environment must also be taken into account and hence we define a social and organisational setting where the conditions and circumstances are to some extent similar. The workplace is one such delimited and manageable setting, subject to the important insight that workplaces differ. When workplaces are compared, their differences as regards conditions probably exceed their similarities. A standardised program (*one size fits all*) does not work. It is the unique circumstances of each workplace which decides what the work of health promotion is to involve and how it is to be carried out. It is therefore necessary to apply what organisation theorists call the *situational way of looking at things (contingency theory)*. This implies that principles for organisation, planning, management and leadership, cannot be automatically transferred from one workplace to another. Instead of generalisation, emphasis is placed on the fact that organisational planning and leadership must be based upon the actual context and situation, with attention being paid to the unique conditions which apply to the workplace in question, at all times. This viewpoint applies in particular to internal issues of the organisation such as working environment, management and the status of employees. In accordance with this way of looking at things, Bolman and Deal[179] hold that the task of organisational research is to develop analytical methods and use several perspectives which allow us to describe the current state and to apply various organisational principles, depending upon the change to be carried out.

There are also external circumstances which distinguish organisations from one another. There is, for example, a crucial difference between the control mechanisms at work in public and private enterprise. The principals, who steer organisations in the public sector, are public authorities - state authorities, county authorities or municipal authorities-where the task of the employees is administration or the provision of social utility. Control occurs

Chapter 10

via legislation, regulations, budgetary methods and frameworks. These circumstances also influence the conditions for organisation, management and development, with the objective of creating workplaces which are better designed to promote health.

In private enterprise, it is in general the market which steers. The enterprise can develop freely in relation to what the company is able to produce and what the customers wish to buy. The employees can thus participate more directly in the economic success of the enterprise. Often - for good or ill- the financial performance is the goal of the enterprise and determines how the company organisation deals with issues affecting staff and health. Economic performance can create possibilities and resources for a good staff policy. Economic considerations and decision-making based on them can also mean that the work is aimed more at intensity and quick results, than at human conditions and sustainable activity.

Hancke[180] hold that there are differences between conditions for intensive work systems (IWS) and conditions for "sustainable work systems" (SWS). The degree of intensity is based on institutional differences which, in their turn, create conditions for how leadership is exercised and which also supply employees with different resources for negotiating about, and changing the way the work is done.

In general terms, the workplace is a good setting for health promotion when working life is organised and has a comparatively clear-cut structure with given frameworks or limits. There are rules and routines for decision-making, communication, work processes etc. Above all, there is in working life a great deal of knowledge and a long tradition of working with development and change.

The majority of workplaces ought, therefore, to have a good potential to both host a health promotion process and, as a result of that, to be a supportive environment, that is to say, an environment which contributes to health by the interaction between individual

The workplace as setting

people's attitudes and behaviour and the structural and functional conditions which are unique to the workplace in question.

System viewpoint

The concept of workplace is a general term for the setting where people are present and subject to certain shared conditions. One characteristic is that in it some organised activity is carried out with the aim of achieving a result or attaining a certain goal. The workplace has also an organisational structure and a collection of people with specific roles.

In the real world, it can be difficult to define what a *workplace*, a *work group* or an *organisation* is. Workplaces can take many different forms and consist of a varying degree of complexity and a varying number of sub-processes and sub-systems. It is necessary in health work, based on the idea of setting, to clarify what the setting is, and decide what falls within the setting and which individuals are to be considered part of it. In certain companies, the work can take place on one particular site, whereas in others it may be carried out at several separate locations and on occasion, the organisation may be completely virtual[181].

When responsible for the planning and management of a health promotion program, just as in the case of any development work whatsoever, you must map out at an early stage "the landscape" of your setting. The larger the organisation, the more complex the map will be. A small company with ten employees has a much simpler map than a large company with "line organisation" and many levels.

In dealing with the complex setting represented by an organisation or workplace, it is of great help to use a *system viewpoint*. As mentioned in chapter 7, the system viewpoint[182] can be seen as a language or conceptual apparatus which can be used to describe

Chapter 10

various phenomena such as organisations, people and families, right up to large settings such as companies or the whole of society. This viewpoint helps us to understand the links between parts or levels, in different systems or between different systems. It has been shown to be particularly valuable to use this approach to study hierarchies of systems (e.g. an individual- a group- a department- a factory- a company etc). The level A lying immediately above a given level B constitutes a *metalevel* of B. (See Fig. 10.1).

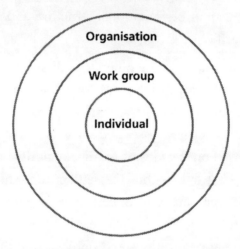

FIG. 10.1 *System levels of organisation*

There are several theoretical applications within system theory which can contribute to understanding how organisations and workplaces function. Johansson[183] describes Scott's three basic models for the system view of an organisation:

1. The organisation is defined as a rational system which is characterised by division of labour, a strongly formal social structure, rules, principles, *chain of command* etc which would seem to be able to be regulated by a chart of the organisation. This organisational structure is best illustrated in accordance with the tradition represented by the school of Scientific Management[184].

2. The organisation is defined as a natural system, where the organisation is likened to a collective whose participants are only to a very minor extent steered by formal structures and official targets. Instead, the system is kept together by a common desire to preserve the system. The system is a goal in itself. Informal relations, roles and norms are evolved by the people who are active in the organisation. A metaphor which illustrates this organisational structure is the biological organism where various processes -providing there is oxygen and energy is generated- work together with the aim of staying alive.

   Ahrenfelt[185] uses this picture as his starting point when he describes the process which can arise in an organisation where there is non-existent or weak leadership. When people lack leadership, their energy is transferred from the collective field i.e. the goals of the organisation, to their own private goals (own survival). Ahrenfelt calls this process of dissolution, *mental entropy*. It is an energy loss in the organisation which weakens and limits its operative capacity.

3. Finally, according to Scott, the organisation can be defined as an *open system*. This definition appears self-evident for organised social systems. Organisations must be defined so that they include the whole of the environment with which they exchange flows. The surrounding environment has a major influence on the structure, activities and performance of the organisation. The integration of the open system with its surrounding implies that every moment and every point in the system constitutes a unique situation.

Chapter 10

To summarise, the system point of view allows us to understand the complexity and negligible predictability of the organisation. It also helps in bringing out the hierarchical levels of the organisation. A common division of the organisation is to divide it into *individual level*, *group level* and *organisation level*. These three levels constitute a first rough map or the minimal number of input levels involved in health promotion work. Even in a small company (with the exception of a single person company) with only one workplace, there are these three levels. On the basis of this division, it is then possible to frame questions about the responsibility of the company or owner, the responsibility of the boss, the responsibility of the joint work group and finally the responsibility of the individual employee with regard to seeing that health is preserved or improved.

Every level in the organisation controls its own specific conditions for work and health. Before the planning of the health promotion work takes place, it is valuable to clarify

- at what level are the various questions decided?
- at what level is a particular thing done?
- what can be influenced at a particular level?
- what goals can health work have at a particular level?

Health promotion work based on the idea of setting should set out to cover all the organisational levels in one way or another. Measures which are directed at the lifestyle of the individual are still of great importance, but they cannot suffice for the whole change in creating a healthy working life. There must be integration with other parts of the organisation, an interaction appropriate to the purpose and interplay between the various subsystems. In planning and preparing the work, ideas and initiatives often come from particular individuals, but integrated health promotion work which is far-sighted and deals with the organisation as a whole, must engage the support of the organisation's top management to push things through, if it is to be successful. Ilona Kickbusch[186] holds that "... *top level commitment is absolutely essential for success*" but she warns simultaneously that "*if the development process is not participatory, it is doomed to failure.*" (p. 345).

Health promotion can be described as a kind of balancing act between a "*top-down*" initiative and a "*bottom-up*" commitment from every individual employee. Top management's visions and policy statements must be counterbalanced by the employees' sense of freedom, possibilities and resources. The energy and motion forward in a process of change must come in large part from the operative health promotion work, where the organisation's work groups and management at different levels are involved. Workplace meetings, management meetings, and meetings with employees are occasions when visions about health in the workplace can be communicated and made concrete. The will to participate is, in the end, largely determined by whether each particular individual judges it meaningful to participate and become involved.

Health promotion is a process which aims at better conditions for the individual's health. But when it is done properly, it can also be an effective strategy in the management's task and goal of influencing the organisation's capacity to carry out its mission. A management which has seen the link between health and productivity has also acquired a powerful motive for giving this form of health work scope and resources.

Integration

In the Luxembourg declaration[187] emphasis is placed on the fact that health work cannot be an activity on its own divorced from the rest of the organisation. It must be integrated with the rest of the activities: *Workplace Health Promotion has to be integrated in all*

Chapter 10

important decisions and in all areas of organisation s(integration). (p.2)

This conclusion is especially obvious and necessary in the case of health promotion, since this form of health work includes and emphasises the surrounding environment. Most of what happens and is done in a workplace, is more or less linked to the health of the workforce, and for that reason, the health issue deserves to be part of the daily routine in the majority of contexts where the organisation has to make a decision. In certain organisations, this can be difficult to achieve, since the tradition that health is a matter for the individual, for the safety representative, or for company health care, can be deeply rooted in the organisational culture.

The customary division into organisational levels is that of individual, group and organisation. This is most often a natural and practical structure to use as a framework for describing what is happening in various organisational contexts. An analysis of the current state, or a plan of action, presupposes decisions, resources and the assuming of responsibility which can be more easily traced back to the right level, if this division is adopted. As an illustration, one can point to a major pharmaceutical company which has a tripartite division of the responsibility for health in the workplace, namely the employee's responsibility, the senior management's responsibility and the organisation's responsibility. As a result, all the various initiatives against ill-health and improved health are assigned a certain position in the hierarchy of responsibility from a decision-making and operational standpoint.

Many issues, which are important for health in the workplace, cannot be arranged so that they can be located at a specific decision-making level. There are activities and processes which affect several organisational levels and their importance from a health perspective must be judged according to their own particular circumstances.

The workplace as setting

A physical question relating to the working environment can, for example, be discussed by the line manager and safety representative, but require a decision from a higher level in the hierarchy. At the same time, at a practical level it may be dealt with by other employees who are lower down the chain of command.

As the work of health promotion becomes all the more integrated in the organisation, and awareness of health issues grows, more and more of the ordinary activities will be shown to have a connection with health. The following examples show how the integration of health work with workplace's life and activities can take place.

- integrate health in decision-making
- integrate health in policies
- integrate health in planning activities and goals
- integrate health in organisation
- integrate health in the development of human skills and knowledge
- integrate health in management and leadership
- integrate health in quality work
- integrate health in workplace meetings
- integrate health in one-to-one management/staff personal interviews

Integrate health in decision-making

In daily decision-making, there are many issues which have, - or can have - an effect on people's health. The staff in a workplace are influenced positively or negatively, for example, by how issues relating to organisation, the way the work is managed and directed, ergonomics and other questions concerning the working environment, are dealt with. Even here, a system approach can be

Chapter 10

applied and show that there are factors and decisions which affect people throughout the organisation: top management, the work team and the individual. In the traditional settings approach, the external influence of the company on its surroundings, local society and the external environment are also taken account of.

Organisational priorities and decisions can sometimes involve conflicts of interest where alternatives such as social utility, utility for the owner, commercial utility and utility for the employees conflict. The majority of organisations -and not simply commercial enterprises - have to take note of the economic aspect as well as of productivity, and these probably still constitute the most common priorities in decision-making.

The idea of health promotion does not deny the importance of economic targets, but it aligns itself with the ideas of researchers such as Mari Kira[188] who holds that the long term sustainability of groups, organisations and societies is based on individuals' well-being:

Long term sustainability at group, organizational and society levels builds on the well-being and development of individuals and their continuous ability to face new challenges and deal with them alone and collectively...Sustainable work systems, by definition, do not make trade-offs between organisational results and people. (p.29).

Integrate health in policies

Work environment and the health issue involve a great number of value judgements, as well as organisational policy. The policy statements and other directives concerning personnel and health issues which the leaders of the company or administration circulate, should contain core values and describe the priority assigned to these issues within the organisation. The management or - in the case of a public body- the principals- can win the personnel's trust and involvement by formulating and publicising values which give priority to human issues. According to Collins & Porras[189], a policy which benefits the employees, society and the company *in that order*, has shown itself to appeal to people more than a policy which is purely concerned with economic or other performance-related measures.

The policy statements concerning work environment and health work can subsequently also serve as a guide for how the health work of the organisation should be organised. Should we apply three parallel strategies? Which activities are the responsibility of the organisation and which are the responsibility of the employees themselves?

The policy can also embody a "code of conduct" for how management and heads of administration deal with the health issue in their daily work. What criteria apply to management from a health perspective? How should one act as top manager and employee in dealing with specific problems and questions which have to do with work environment and health?

Integrate health in planning activities and goals

If activities are to be well-organised, planning is needed- long term, medium term and short term planning. Visions and goals are set out and operationalised, right down to specific tasks to be achieved by individuals. In many contexts, there can be reason in planning to take into account various health aspects. The question is whether the central processes[190] of the activity can be designed in such a way as to reduce the risk of ill-health and promote those conditions conducive to the preservation of health. Can this be

Chapter 10

accomplished in a way which is simultaneously positive for the organisation's performance and achievement of its targets?

Obviously, health work in itself can also have a place in the plan for the enterprise and be as much the object of serious planning as the rest of its activities. The organisation which starts up a health promotion project, is initiating a process of long term change. The change envisaged must have both goals and plans in order to be implemented efficiently. In the planning work, questions about goals, structure, resources and the scheduling of operations must constantly be balanced against the enterprise's central processes and against the non-negotiable demand for general acceptance by the parties involved, and their participation. Health promotion is a *participative* strategy which must incorporate both order and a certain amount of chaos.

Integrate health in organisation

How should we organise an enterprise so that it will be both efficient and simultaneously promote the health of the workforce? Frederick W Taylor[191], who published his main organisational principles around 1910, had already thought about the kinds of psychological experiences which certain types of work organisations produce among the employees. His organisational ideas would seem, at least in their application, to have created more discomfort and alienation than well-being and health. His reflections, however, were followed by scores of other researchers who have subsequently studied and noted how the way work is planned, has great importance for human well-being and health. We cannot go further into this large field here, but we shall content ourselves with remarking that these ideas form, even today, what in some ways, is the key issue in the theory of organisation. An efficiently organised activity is suboptimal from a production standpoint, if the employees engaged in it do not have a positive experience of their work situation.

The workplace as setting

The choice of organisational form and even more the choice of a production form, also has repercussions on the physical environment's effect on health. The Swedish automobile industry's attempt in the 1980s to replace the assembly line with work stations and multifunctional teams, aimed at improving health conditions for the employees. Nevertheless this approach to organisation and production failed to deliver the human-friendly automobile factory of the future. Instead, the scheme was aborted, since demands relating to efficiency and rationalisation failed to be fulfilled. During a conversation about working conditions and health at the end of the 1990s, the then Director of Production at one of our Swedish automobile factories, asserted that the rebuilding of the factories at Kalmar, Malmö and Uddevalla in connection with the change to non-assembly line production, was one of Sweden's most expensive and disastrous, industrial experiments. The new lines of production which were subsequently built in Sweden, were forced to re-introduce the assembly line, since competing automobile factories in other countries which were comparable as far as technical quality was concerned, had lower costs in terms of wages and, above all, a queue of would-be employees keen to enter the factory "and work their shirts off at the assembly line".

This does not mean, however, that efficient production cannot mean both health and job satisfaction: once more, it is the total situation which decides the conditions for health. Whitelaw[192] holds that e.g. in development work and organisational change, health promotion can be an indirect motor which contributes to success. Employees who feel well and are able to cope physically and mentally, can deal with a demanding process of change better than others.

Chapter 10

Integrate health in the development of competence

Health and competence (the successful acquisition of human theoretical and practical skills facilitating the solution of problems in a given field) are clearly linked in two ways in working life. First of all, there is the importance that peoples state of health has for their capacity to carry out their tasks at work. The capacity to be adequately productive, in the way that is expected, is determined, both by how well people feel, and by how far their bodies are up to the task in hand. In 1998-1999, a project *Ett friskt arbetsliv*[193] [*A healthy working life*] was carried out under the direction of the Royal Swedish Academy of Engineering Science [IVA]. In connection with it, representatives of commerce and industry emphasised the importance of clarifying the connection between the company's efforts to create value and the competence of its employees. The authors took the view that the ability to draw upon the competence of the employees and a sustainable working capability are in turn dependent on good working conditions and good health.

According to Ellström's[194] definition, it appears that competence is dependent upon a spectrum of human qualities. Competence is:

> "an individual's potential capacity for action in relation to a certain task, situation or work. This individual capacity to act can, depending on the character of the task, refer both to knowledge, intellectual and manual skills [färdigheter] and to social skills, attitudes and personality related characteristics in the individual." (p.149)

Well-being and health status form in this connection one of the keys to competence and the exploitation of competence. Health is seen to form part of the competence profile and thus ought to have a status as a value-creating resource in working life and be treated as such.

Given such a conclusion, the next step follows clearly. Knowledge about the conditions governing people's health ought to be adequately recognised in the organisation. Every employee ought to be aware of the factors influencing health and how they operate. It is not simply a question of knowing about risks linked to the working environment, but also knowing about how we can improve our health. Since health is "a perishable good", health work at individual level is a constant work of improvement. For top management and other key personnel who have the possibility of influencing things in a major way, there are reasons other than purely individual ones, why knowledge about the conditions governing health is included as a self-evident part of competence development.

Integrate health in management and leadership

Health promoting leadership is not essentially different from good leadership in general. The importance of leadership for the work environment has, nevertheless, been a recurring question in discussions for some time. On occasions, top management have become the scapegoats and even considered as the most important reason, both for a bad working environment and a high rate of absence due to illness. Such a picture is many times totally misleading and unnecessary. An evaluation of the effects of leadership can only be carried out, if we simultaneously describe the conditions applying to leadership. One major reason for difficulties in leadership, we see in Swedish working life today, is to be found in the change to so-called "lean and flat" organisations which took place from the 1980s onwards. Many managers have disappeared due to streamlining and rationalisation, which has meant that there are workplaces where the manager is responsible for up to 100 persons and more, and that is too much to make good leadership possible.

Chapter 10

A supportive personal leadership which is good both for the manager's own health and for the health of the employees, in his or her charge, presupposes that the work group has an upper limit of 20 or so people.

When we employ the term *health promoting leadership*, three different aspects of the managerial task can be meant.

1. *Personal leadership* in the relation between the manager and staff can be important for health. The job of management is to support, encourage and preserve the link with the individual employee. There are several other qualities such as being clear and lucid, setting goals, listening, communicating and holding constructive personal interviews with individual employees which also help to characterise health promoting leadership.

2. *Tactical leadership* (sometimes referred to as *pedagogic leadership*) is a dimension of health promoting leadership. The relationship between the manager, the employees and task should function in a way that will benefit all of them. Goals will be achieved, people will feel good and the manager will personally experience a sense of job satisfaction. The tactical role is a pedagogic leadership role which, in the wisest way possible, should represent and combine the organisation's task and goal with the potential and needs of each employee. Maltén[195] describes pedagogic leadership as an application of five leadership dimensions aiming at goals, relations, innovation, situation and ethics.

3. *Strategic health promoting leadership* is the third meaning of health promoting leadership. This is the manager who consciously and purposefully works to develop the workplace so that it becomes a health promotion setting

The workplace as setting

by pushing forward with the health promotion work both strategically and operationally. He or she helps to involve and stimulate the workforce in what is being done and integrates it with the other management systems of the organisation and with the technical, environmental and social aspects. This kind of manager is also deeply involved in the health issue as it relates to his or her own workforce

Today, more and more emphasis is placed on the importance of the management of a company integrating health issues with productivity. In several countries, this subject is discussed under the heading of *Health and Productivity Management* (HPM)[196].

Integrate health in quality work

There are many views about how we should deal with quality work. An increasing number of workplaces allow themselves to be certified according to some form of quality system. Total quality management (TQM), for example, is a way of organising the work in hand and thereby raising productivity and profitability.

For some companies, a quality certificate is necessary in order to be allowed to supply their goods or services. It is often the external, customer-experienced quality which is crucial. The internal quality in the workplace becomes a means of achieving the external quality.

The importance of health for quality can be described in various ways. The logical link ought to be that people, who feel good, are also those who most often are able to maintain high quality in their work. In order to maintain high quality, factors such as clear goals, good organisation, an assignment of work that is reasonable and a good working environment are needed. Because the same factors also give the employee a positive experience, this means that health and quality management can co-operate in a positive spiral of improvement.

Integrate health in workplace meetings

The process of democratisation in Swedish working life has meant that we have acquired new forms for dialogue and decision-making. Today, the majority of work groups have regular meetings at which information is distributed, common problems are addressed, decisions are taken and there are activities to stimulate group solidarity. The value of these meetings and their agenda has become clearer and better with the years.

Today, tasks and roles at work are less enduring and predictable. Instead, they must constantly be adapted to new conditions in the organisation and in the world around. At the same time, the authority to make decisions- with the associated responsibility- has been moved downwards through the organisational pyramid to work groups and individual employees. This has meant an increased need for dialogue and co-operative planning. The manager, together with his or her team, needs to size up the situation, clarify what can be done, given the possibilities at hand, and plan for the immediate future. The workplace meeting should not only deal with questions about the activity carried on in the workplace; it should also provide an important forum for discussions about work satisfaction, well-being and health. Many workplaces today include "work environment and health" or similar topics, as part of the regular agenda at their meetings. Moreover the actual way of running these meetings is an important instrument for influencing the organisational culture and spreading those values which have the greatest importance for job satisfaction and health of the employees.

Whitelaw et al.[197] hold that *participation, the development of empowerment* and *ownership of change* are important ingredients in the creation of a sustainable change to better conditions for the health of the employees. It is this type of participation-creating effects which well-run workplace meetings can contribute to.

Integrate health in one-to-one management/ staff personal interviews

In many workplaces in Sweden, the one-to-one management/ staff personal interview, or development interview, is an established instrument in the manager's toolkit of leadership. Here the health issue also plays an important role. At such one-to-one meetings, the manager is able, not only to speak about work and the development of competence, but also to gain an understanding of how the employee feels and copes with his/her work. A manager, who is interested in health, can also be a health coach who gives tips and encourages employees to preserve their physical and mental capacity to cope and their satisfaction in their work.

Because health promotion work strives for integration, this does not mean that everything that happens in a workplace should be called health work. It is more a matter of introducing an awareness and a way of thinking about the causes of health, about its effects in a workplace and about its importance for people feeling well and functioning well. As a result of this extended view of health, the workplace employees also have a better chance of changing and developing the work being done. Old well known activities are seen in a new light and new ones are created which benefit both people's health and the performance of the company. Workplaces are made up of 'whole people' (i.e. people who interface with reality in a multitude of ways) and the health question thus also requires an integrated and holistic approach.

Chapter 10

Differing preconditions for health promotion

The settings approach is a way of theoretically structuring and defining the boundaries of health promotion work. In real life, the responsible health worker or whoever is in charge of the initiative, must always consider, on each occasion, whether health promotion is a suitable strategy, given the situation which prevails in the specific workplace. If the program cannot be implemented in its entirety, it is a question of deciding the extent to which these ideas can be incorporated in the organisation's own activities, or the way in which a more ambitious long term introduction of health promotion can begin in small steps. It is extremely demanding when a health perspective has to permeate every level of an organisation. All organisations do not have initially the culture, communicational ability or competence which is required for a process of health promotion. Different organisations can be said to find themselves at different positions on a scale from complete passivity and more or less zero capacity for change to those which have a powerful capacity for change and development and where discussion and dialogue about values and the future are welcomed.

Whitelaw [198] chooses to describe the relation between the setting's "nature" and the form health work can take in various organisations, as two extremes. The table below, due to Kylén[199], characterises workplaces in terms of defensive/offensive routines. If we call the two extremes the immature and passive workplace and the mature and active workplace, we have the following figure:

The workplace as setting

	The immature and passive workplace ⟷	**The mature and active workplace**
Description of the nature of the setting	Hierarchical structure Centralised culture Power according to position Central control ⟷	Flexible open structure Decentralised culture Learning stimulated Delegation of responsibility
Group's behavioural pattern	Internal adaptation Someone else's responsibility Keep to what is given Self reliance Closed communication Parallel work Rigidity Adaptation to rules ⟷	Adaptation to the world around Joint responsibility See possibilities Balance in self-team reliance Open communication Co-operation Flexibility Situational adaptation
View of health work in the organisation	Health is entirely a matter for the individual ⟷	Health is an obvious strategic issue. Promoting health is an important part of development work
View of human beings	Passive recipients of measures ⟷	Active in creating their own work environment
Type of health work	Directed at measures with limited initiatives according to needs and legislation ⟷	Long term health promotion strategy, integrated at individual, group and organisational level. Supplement by rehabilitation and preventive initiatives
Health worker's role	Traditional profession-based services ⟷	The health worker, in addition to his basic profession, is also a process leader who co-ordinates external and internal resources-teamwork.

FIG. 10.2 *The "immature" and "mature" workplace*

Chapter 10

Working with health promotion as a strategy is, among other things, about applying an approach geared to change which consists of gradually, by means of organisational learning, influencing such things as understanding, culture and co-operation in the organisation. Bjerlöv[200] describes this successive learning and changing state as a dialectical process where the changed views of individuals eventually mean that the organisation moves into a new state which in itself influences individuals to develop further. This type of learning process requires both time and structure. The people in charge of such work need to be well-versed in the skills required for orchestrating the processes of change and development.

Summary

The workplace as setting is both a life environment which influences health and a context- a setting for the work of health promotion

Health promotion has a primary role as a model or strategy for work directed at improving health in the workplace. At the same time health promotion, as a democratic and integrated process, has the possibility of being a driving force in overall organisational development, where the health perspective serves as a resource for achieving organisational goals.

A system approach in which the workplace is divided into various decision-making and operational levels e.g. individual level- group level - organisational level, can help to assign responsibility and activities appropriately. Integration and the link between levels is necessary in order to find and implement solutions in many issues.

Every organisation and workplace has its own unique potential preconditions for health promotion. Size, maturity and culture decide if and how a health promotion strategy is to be applied. There is also, to a varying extent, a whole number of interested parties who affect or are affected by what happens in the workplace. There can be many people who want to be included and have something to say when decisions are taken about the health work. There are established actors who have worked practically with health work and who wish to have as prominent a role as possible. These established functions and professions form a resource which should be made use of. However no group, by itself, is the self-evident principal actor in the health promotion approach.

Health promotion is an upgrade to an approach that adopts an integrated process-oriented way of thinking, rather than one that thinks in terms of profession-based responses. This also means organisational innovation such as team work in interdisciplinary groups to which people are recruited more on the basis of the task in hand and the goal, rather than on the basis of professional title. Perhaps the most important, and at the same time, the most difficult task is to create forms for participation and working together which lead to efficiency from a work perspective and simultaneously satisfy the psychologically important need for each and every one who is affected , to feel that they are significant.

11 Conditions governing participation

The most frequently occurring and most crucial criterion of health promotion is that people themselves must be in a position to be able to actively participate in health work. This is a radical requirement, in the sense that we go from an *expert perspective* where some specialist has the knowledge, to an *actor perspective* which entails that we treat the individual human being's knowledge and empowerment, that is to say their power over their own situation, as perhaps the most important prerequisite in promoting health. However, it ought to be self-evident that when we consider health conditions as part of a social setting, the people who make up that setting must both want and be able to participate, if any change is to come about.

The concept of participation within health promotion

In the documents which WHO has drawn up through the years, participation has been a crucial requirement for creating the prerequisites for health. Already in the interim constitutional document of 1946[201], the importance of co-operation and participation for the general public and the individual was emphasised. *"Informed opinion and active co-operation on the part of the public are of the utmost importance in the improvement of the health of the people."* (p. 100). Since then, the condition relating to participation has been part of the majority of documents which have been published by both WHO and ENWHP on Workplace Health Promotion.

Chapter 11

The description of the importance and meaning of participation varies in the literature and can be said to be coloured by two value-based perspectives. The first consists of documents which adopt a more traditional, patriarchal and formative attitude. According to this view, human beings are the objects which form the focus of health promotion work and accordingly they ought to take part and have opinions about how this work should be carried out. Often without being explicitly said, it is simultaneously considered a good way to make people motivated to change their behaviour.

The second perspective looks upon human beings as the subjects in health promotion work. The participation, which is aimed at, proceeds from individuals themselves, by allowing them to be empowered so that they themselves can actively maintain and develop their degrees of freedom. The change which ought to take place perhaps depends on the surroundings and system much more than on their own habits and behaviour. In a workplace, each individual is part of a system with certain given conditions which impose limitations on individual freedom. These conditions must be balanced against goals such as autonomy, self-empowerment and integrity. Decisions occur in co-ordination with others and individual attitudes and positions must find a balance between self-interest and solidarity. In 2002[202], WHO published a document which contains criteria for Workplace Health Promotion with a clear actor perspective and broad view about what participation can be:

For the successful development of workplace health promotion management, it is important:

- to recognise the central role of the empowerment of employees, in terms of competency, level of autonomy, and sense of coherence.

Conditions governing participation

- to ensure an appropriate balance between the processes of effectiveness increasing and the capacities of the workforce.
- to include a comprehensive understanding of health in company policies and in all procedures involved in a continuous improvement process
- to identify factors contributing to development of health
- to facilitate and strengthen impact of such factors conducive to health of all staff
- to ensure the establishment of an enterprise-wide participatory infrastructure; and
- to enable all levels of employees to share their interests and expertise with the key players. (p.27)

In this passage, we find an emphasis on factors creating participation such as *empowerment, autonomy, sense of coherence, balance between demands and resources, participatory infrastructure* and the possibility for the workforce at all levels to take part in a *dialogue* about health.

The concept of participation in working life

The organisational ideas of Taylorism, which we mentioned earlier, were based on a rigid structuring of the individual employee's tasks without individuals themselves being able to participate. Such a Tayloristic, functional and technical organisation is still operating in the twenty-first century in branches where rational production and high efficiency form the principal goals. At the same time, from the beginning of the twentieth century onwards, there has been a trend towards introducing more and more participatory processes to replace those relying on the detailed management and organisation of tasks. Through the years, one can find in the literature a host of examples of this. In this connection, it is especially worthwhile mentioning the interest in organisational learning which began in the 1970s with researchers such as Bateson[203] and Argyris & Schön[204]. According to their viewpoint, the role and participation of the employee is crucial in developing knowledge of how the different parts of the organisational system can be made to function in concert with one another in an optimal manner.

Creating participation can be a way of empowering the employees so that they have control over their own situation at work. Karasek and Theorell[205] broke new ground in describing the connection between demand and control and how this in turn affected individuals' subjective experience of their work situation. The concept of control is defined differently by different researchers, but it is, among other things, a question of how much influence individuals have over their work and in their work. Influence means being able to affect the conditions determining *that* the work is carried out, as well as *how* it is carried out. Influence is being able to make one's own decisions and being allowed to participate in collective decision-making in the workplace.

In workplace health promotion work, it is possible to distinguish two opposite tendencies about where in the organisation we should place the focus and responsibility for the health initiative. In the first, the focus is shifted from the individual to the organisation. Up to the close of the 1990s, great attention was devoted to the role of individuals, their lifestyles and the responsibility they had for living in a healthy manner. Employers have looked upon their responsibility as being one of encouraging fitness & wellness programs and influencing individuals to accept with gratitude the invitations to courses geared to training and stress-release. The focus has, at least in Sweden, steadily moved away from individuals, first to the role and responsibility of the workplace for working conditions

Conditions governing participation

and health, and subsequently to the role and responsibility of the organisation. Concepts such as the good workplace[206], sustainable work systems (SWS)[207], sustainable working life[208], a healthy working life[209] and quality of working life (QWL)[210] are used in the literature as headings for health-directed work aiming at change within a larger entity.

Simultaneously with this shift of responsibility from the individual to the organisation, there is an opposite trend as regards work, in general, concerned with change and improvement. It has become more and more customary to stress the employee's role in ensuring that change takes place. The shift of focus from the organisation's management to the individual is linked to the realisation that changes in attitudes and behaviour rest on the fact that the people concerned are encouraged to participate. In practice, these trends hopefully mean that there will be possible to see more of a balance between individual issues and organisational issues and a balance in sharing responsibility between employees and employers. With its approach in terms of settings, health promotion adopts a holistic view of the workplace and strives to achieve a balance in the initiatives between the various system levels all the way down from the whole organisation on the one hand, to the individual on the other.

Why participation?

Participation is a word with a positive connotation and an obvious ethical basis and political association. Everyone should take part and enjoy equal opportunities. There are also more practical reasons why participation is central to health promotion, and why anyone participating in any kind of organisational work aimed at change, ought to reflect about the significance of participation.

We shall examine four motives for participation.

- Participation is a way of pooling the *knowledge* of several people.
- Participation is a way of creating a sense of *involvement*.
- Participation provides a better *adaptation to the situation*. Greater chance of acting wisely and correctly.
- Participation is essentially *regenerative*.

When we are making decisions and planning for change, what serves as the basis of our *knowledge*? Is there additional, quite different knowledge about how our company works than that which I, as a manager, have inside my head?

"Why should one involve the whole workforce in the process of change? It can lead to a lot of bother and take much longer than if we present a finished plan." Such a pronouncement is perhaps unusual at the beginning of the twenty-first century, but such thoughts are still to be found. It is an approach which can depend on a shortage of resources in the organisation or upon uncertainty or lack of insight on the part of the management.

An organisation consists of people, all of whom possess a brain. If a contemporary organisation still thinks in Tayloristic terms regarding the division of labour and detailed management of each task, it is presumably rather convenient to opt for the first model where the management draws up the plan entirely on their own. However, a process of change and development which is based on such a strategy fails to take account of several important levels of competence and value-creating labour inputs. All the people in a workplace have knowledge to contribute and this intellectual capital is a resource which should be used. Knowledge and understanding of the reality can also, in favourable circumstances, be developed to a greater extent in dialogue with others. Participation gives a chance to reach a collective view about the prevailing situation and

Conditions governing participation

the purpose of the change, as well as an understanding about how best to carry out the work. When there is a shared understanding, decisions are more broadly based and possess greater legitimacy. The strategic priorities and decisions which rest with the management, can also be more secure when time is taken to "sound things out" and the proposed process of change has become better known. It is obvious that the preconditions for such an approach vary widely. The way the work is organised and the maturity of the workforce as regards such a participatory strategy, decide the manner and extent of the process.

From time to time, consultants are used as experts in processes of change. As an expert, there is always a risk of falling into the trap of considering one's own well documented solutions as the best there are in the majority of situations. The management group who receives this expert solution finds itself assigned a subsidiary role and as a result, not enough attention is paid in the planning to their own ideas or to the workforce's experience. This professional dominance has in most situations much to win from a reorientation which allows for dialogue and listening to what the people with the problem know and have to say about their situation. Sandberg and Targama[211] hold that there is really only one way to understanding-based knowledge in social systems such as a workplace. Every person develops their understanding in conjunction with other people via dialogue and participatory processes. Understanding is a relationship and a way of responding to phenomena in the situation in which I find myself. In order to increase both my own and our joint understanding, we must participate in this relationship. Both the manager vis-à-vis his workforce and the consultant vis-à-vis the client need to have a supportive and consultative role in this process. Expert knowledge can be needed to act as a director or moderator in the participatory dialogue.

Chapter 11

The second reason for encouraging participation is the value of having access to people's sense of *involvement*, the energy which feeds the process of change. Within human psychology, there is great emphasis on the persons undergoing the experience and their subjective interpretation of what is going on, which also becomes their version of the truth. Alvesson[a] described this as the opposite of authoritarian thinking, where someone with higher status knows better than the individual concerned. Human beings wish to be creative; they make their valuations and choices on the basis of what is meaningful according to their own view. The ready-made, mechanical and authoritarian solution to a problem provides neither stimulation nor motivation. Already in the 1950s, Herzberg[213] launched a theory about what motivates people in their work. He distinguished between internal and external motivational factors. The internal factors involve the interaction in the workplace and the experience of our own role in relation to the tasks in hand. According to Herzberg, feeling a sense of collective responsibility, being allowed to work with tasks carrying status, succeeding with something and gaining positive feedback from workmates and management etc-all have great importance for motivation.

Drawing on this, we could adopt a purely instrumental attitude and state that participation is good for psychological reasons. People become more motivated when their opinions are asked for, when they can make suggestions and in other ways feel that they have some importance and status. It is a long-established truth in occupational and organisational psychology when one is studying processes of change and development, that the earlier the workforce becomes involved, the less will be the opposition to the change. Instead there is better adaptation and more purposeful energy is put into the change[214].

An application of this view stressing involvement is to be found in those managerial theories which instead of external control via instructions and control, apply internal self- control where it is people's own view of the work's goal and meaning which drives them. This motivation is particularly strong when people themselves participate in the formulation of goals and visions. Examples of this approach appear under such titles as visionary leadership, idea-based leadership[215] and transformative leadership[216]. Irrespective of what we call it, there is also here, as Trollestad[217] among others, has pointed out, an instrumental view about human beings' importance for work and organisation. This places demands on the person in charge of the process of change to reflect about the conflict of interest between the autonomy of the individual and organisational efficiency. It is easy to say 'participation' and speak about every employee's independent and free choice, but it is sometimes difficult to apply this in a way which is equally useful to both the individual and the organisation.

The third reason for increased participation is the possibility of better *adaptation to the situation*. The actual workplace has its own particular conditions and those who work there are best placed to know what they are. Every organisation in each situation constitutes a unique context or set of circumstances which requires both the knowledge as it is experienced by the employee, the manager's overview and the expert's general experiences, in order to find its way correctly.

This way of looking at things ought to be just as useful when we want to create better conditions for health in a workplace, as it is when the organisation has to be adapted to the market and the world around. Given the complex nature and circumstances of a workplace, it ought to be completely obvious that a transfer of tailor-made solutions from one workplace to another is bound to be a risky business. Such a procedure easily loses sight of the possibility of reaching a joint understanding of why a health project ought to be implemented and runs the risk of leading to more resistance to the

Chapter 11

imported ideas than to goodwill and involvement. It is seldom the case that a workplace stands and waits to start yet another project, when it has already enough to do. For that reason, it is necessary to describe what one is setting out to do, so that people can recognise themselves in the description. In other words the project is carried out with their own personal involvement. The scale of the project, its timing, its integration with other processes and much else have to be taken into account.

Moldaschl and Brödner[218] -as we have mentioned earlier- employ the concept of 'reflexive' intervention to describe a strategy for implementing change where one chooses participation, dialogue and reflection to achieve legitimacy for the change envisaged, and in order to be able to make use of the workplace's unique history, culture and context in the process of change. They formulate this as follows:

> To follow a reflexive methodology of intervention means to remain sceptical of any universalism, any recommendation of "one-best-way" or "best-practice". It means accepting that any social system has its history, culture and context (e.g. Bourdieu 1990) and must be treated according to that specificity. Instead of simply recommending or transferring knowledge and solutions from other organisations to a present one, the reflexive researcher or consultant would see his or her main task in the contextualization of experiences for the specific case. Instead of offering general problem solving models, the researcher would prefer a thinking in dilemmas- such as between stability and change, autonomy and responsibility, involvement and relief. (p. 185)

We do not exclude the general experiences of the consultant or researcher completely, but hold that these must be reflected and used in a reflexive way.

The fourth reason for participation is that in itself it can contribute to meaningfulness and *regeneration*. The idea behind salutogenic health promotion is to focus on the conditions which contribute to health and on the basis of the four criteria or critical components, create a movement or change to bring about these conditions. *Participation* is a criterion which helps to bring about this movement, but it is also in itself health-promoting. This is closely linked to what other researchers call *regenerative* activities. Jan Forslin[219] has written about this and holds that, hitherto, ideas about how good work can be created, has been based on the removal of destructive elements and through devising conditions of work which provide motivation. Forslin proposes an approach to the study of work which goes farther than the hitherto dominant preoccupation with "fixing what is worst" and studying work as a regenerative process which can re-create human energy. (p.45) He puts forward a number of aspects of the human being which are closely linked to the concept of participation and can be assumed to have importance in this connection: *experiencing meaningfulness, feeling oneself to be competent, receiving social support, receiving external confirmation, being allowed to exercise a position where one can guide others, having qualitative meetings* and *having time for reflection and recovery*. Forslin points out that regeneration is important for both social and economic goals. Traditionally organisational research has been dominated by the economic approach, in other words what is profitable for the organisation.

Of course we must not forget that the overarching purpose of the organisation is to achieve results, often measured in economic terms. There is therefore reason to stress the importance of dealing with the economic dimension at the same time as dealing with the human dimension. Economics and efficiency are one side of the coin, while things like ethics, aesthetics, empathy and emotion are the other. .

Creating participation is thus a criterion in health promotion work, but it also has a direct value as a contribution to making daily work constructive and regenerative. Existential well-being is closely linked to participation and social relationships. Perhaps one of the most meaningful things there is, is to be part of a relationship, a human encounter or a dialogue. When these function well, one finds perhaps the strongest experience of being a human being. Egon Rommedal[220] put it in the following way: " There is a possibility and a collective responsibility that every encounter has made us a little more of a human being."

Participation is important for other health related concepts and qualities such as identity, self confidence, self-image, security, pride and several others which we will not further discuss here.

What is participation?

Participation means to *play an active part*. In legal terms, participation also implied *co-responsibility* which has great importance in a settings approach where the different system levels in an organisation have different roles and different responsibilities associated with them. In order for a health promotion process to be successful, all the interested parties or employees must be prepared to take their responsibility. Popularly expressed, there are both rights and duties as well as opportunities for people belonging to an organisation.

In practice and in the literature, there are several meanings and applications of the concept of participation which help both to understand its meaning and to find ways of applying it. Application is not entirely unproblematic since striving for participation can bring with it political, psychological, psychosocial and pedagogical problems. A good way to start is to reflect about certain concepts which are linked in terms of values, to participation:

- Empowerment
- Autonomy
- Co-operation
- Democracy
- Justice

Empowerment is closely linked to participation when it is a question of being permitted and able to exercise influence and is a question of power and influence.

Since the 1970s, the concept of *empowerment* has been very often used to describe the process or change whereby people increase their influence or power over their existence. It has been of central importance in Health Promotion, above all as an approach within public health work designed to strengthen or create opportunities for the inhabitants of local society to influence the conditions of their life. In the Ottawa Charter[221], emphasis is placed on the strengthening of local initiative:

> Health Promotion works through concrete and effective community action in setting priorities, making decisions, planning strategies and implementing them to achieve better health. At the heart of this process is the empowerment of communities- their ownership and control of their own endeavours and destinies. (p.6)

This involves a political dimension in which participation implies that power in a society or organisation is distributed more on a more equitable basis. Sharing power, knowledge and resources with other people is to give people greater opportunities and greater scope for action. In the literature, *empowerment* is also described as an *emancipatory* process, a liberation of people in a subservient and subsidiary position.

Chapter 11

In literature, we find numerous titles containing the term *empowerment* and often in conjunction with the term *participation*. Starrin[222] holds that, the power aspect involved in the concept of empowerment implies that one actively takes control, whether individually or collectively. It is thus more a question of winning power by one's own efforts than of permission being granted by some superior. Other expressions which, according to Starrin, can help to define empowerment are *grassroots organisation, participation, social support, citizenship, competence* and *autonomy*. Empowerment as a process in itself implies that one identifies and removes those conditions which cause powerlessness, both through formally organised and informal techniques. According to Starrin, it is a way of enabling people to satisfy their own needs, solve their own problems and quite simply decide over their own conditions.

The idea "all power to the people" does not fit into the setting of working life. An organisation has an owner -or in the public sphere is answerable to politicians whether at national, county or municipal level- who must decide and be responsible for overarching issues concerning the enterprise's direction and goal. The organisation's management has then the task of leading the enterprise in accordance with the lines laid down. In the management process, however, it is possible to distribute influence within the organisation. A management that is "on the ball" strives to find a balance between what is the responsibility of management and what is the responsibility of the staff. This entails that the spreading of power, that is to say the distribution of authority and influence over decisions is accomplished in a way which benefits both the workforce and the enterprise. An organisation which does not manage to cope with a balanced distribution of power, but remains stuck with only detailed management from the top down, runs the risk of having employees who experience powerlessness. This can lead to negative

Conditions governing participation

states such as *learned helplessness*[223] and *alienation* [224] that is to say the opposite of power, namely *impotence*.

Dahlström[225] maintains that powerlessness and alienation go together. People who are unable to exert any influence, find themselves facing an experience of meaninglessness, isolation and an absence of value in their work which means that ultimately it is only the hours worked and wages which have any meaning. Such a person seems at the opposite end of the spectrum in relation to the participative and involved employee and what the latter can mean for an employer.

Autonomy means the moral right to be free and independent. At the same time, individuals must function at a social level and be able *to co-operate* with other human beings round about. Psychologists hold that there are two basic tendencies in men and women as social beings. One is the will to *autonomy*, that is to be independent and by my own efforts to rule the world. In order to accomplish this, human beings acquire knowledge of the world around them and power or influence and other resources which will help them to be able to exercise control over their existence. If they fail to acquire these resources, they run the risk of being dependent on others and thereby they are no longer so independent.

While human beings wish to be capable, independent and special, there is simultaneously an opposite thrust to belong, to participate and be part of a context. The desire to belong can apply to a workgroup, an interest group, a family, a company, a country and many other things.

Svedberg[226] holds that human beings bear with them the whole time the question of *being a part of* some greater totality versus *being apart*, and of *belonging* or *being independent*. Both life positions represent a kind of non-existence since we can never be so completely immersed in a relationship to someone else that we

Chapter 11

lose our own identity. There must be room both for independence and for the need to belong.

A great deal of socio-psychological theory maintains that individuals must first discover awareness of themselves and independence in order to co-operate with others. At the same time, co-operation with others helps human beings to increase their self-awareness and influences their self-image and feelings about themselves.

Both *autonomy* and *co-operation* can be satisfied by creating an opportunity for participating. Autonomy is supported by receiving information, knowledge, influence, responsibility and authority. This can be given in a formal way, but it is also strengthened in relations with superiors and other employees. When people are consulted, receive praise and feel themselves confirmed in other ways, this contributes to their positive feeling of being capable. When this feeling of competence is reinforced, the desire to take part and co-operate is correspondingly strengthened. One has something to offer and be responsible for, in sharing one's skills and knowledge. The feeling of belonging and the possibility to co-operate is helped by visible activities which are a prerequisite in a participatory workplace. In such a workplace, we find openness, dialogue, workplace meetings and very often a great deal of good humour.

Participation which strengthens autonomy and co-operation also directly benefits people's health. Starrin and Jönsson[227] have noted that people's experience of status and pride promote health and these can be strengthened by a sense of belonging and participation e.g. in connection with their work. The opposite state is a feeling of shame which can cause ill-health among those, for example, who find themselves 'outsiders' because of involuntary unemployment.

Conditions governing participation

Democracy and *justice* are two concepts which are closely connected and are also linked to participation. Democracy, in its proper sense, means that power derives from the people. In working life, this can be applied in certain situations. When a decision is to be made concerning the planning and carrying out of the work, a democratic method of arriving at a decision can be used either through reaching a consensus or by a majority vote. Things are done according to the will of the majority. Many questions, however, cannot be decided democratically within an organisation. From the outset, it is made clear that the managerial team is responsible for certain fundamental and overarching decisions which the employees are then expected to follow. These managers are seldom appointed democratically, but are selected for their tasks by other members of the management.

All this forms the conditions which determine working life and business enterprise. A principal, an owner, a market etc impose compulsory conditions according to which the organisation thereafter can or cannot apply democratic forms of decision-making. Democracy takes second place to market requirements, the task in hand or legislation. The external rules of play must not, however, be seen as an impediment to greater participation in the workplace.

What does influence in the workplace mean for public health? Can a democratic working life help to provide better and more equally distributed health in society? Töres Theorell[228] has given an account of the growth of workplace democracy in Sweden, which took place from the 1970s onwards. The Co-Determination Act[229] of 1976 marked a resolve to break the employer's one-side right to steer people's work. Researchers supported this and helped to convince the employers of the economic value of increased employee influence from the standpoint of the company. Theorell describes the 1970s and 1980s as the golden years of workplace democracy. In international comparisons, Swedish working life occupied a strong position when

Chapter 11

it came to including democratic routines. Researchers studying working life examined the connection between working conditions and stress. Karasek and Theorell published the book *Healthy Work*[230] with the demand-control- support-model, subsequently much used, which was based on the idea that participation and democracy have great importance for people's health.

Here are some examples of how democratisation made itself felt in individual workplaces:

- Conferring with (negotiating with) the employer became more common
- A trade union representative was given a place on the board of the company
- Employee personal interviews and workplace meetings were introduced
- Conversations about wages with the appropriate management representative were introduced
- Within the automobile industry, efforts were made to try organising assembly in terms of groups/teams in order to increase the workforce's influence over the way their work was organised.

During this period, an increasing number of workplaces set about democratising their work routines. Sometimes this was done more for the sake of appearances than from any aspiration to make use of the possibilities that accompany real influence. Many workplaces have begun to work in teams without thinking about why and how it should be done. Many workplace meetings have been held through the years without the participating employees having felt that such meetings were meaningful or served any useful purpose. This perhaps shows how difficult it is to create greater participation and influence by means of legislation and structural measures. How

Conditions governing participation

are the possibilities for exerting influence to be balanced against individual needs and the practical conditions which prevail in a specific workplace? It is not enough simply to introduce a new form of meeting, since its content is influenced by people's traditional images and attitudes. The latter is, in turn, part of the organisation's culture which alters very slowly. Development however has meant that many workplaces have received training in participation and co-operation. The capacity and maturity needed to work together has been strengthened and this is in itself an important resource in allowing one to continue further with the work of change and development.

Despite the many activities in Swedish working life which helped to introduce more democracy and increased influence during the close of the twentieth century, the development was not entirely positive. Theorell holds that the individual's scope for decision deteriorated, especially at the end of the 1990s. The mental demands increased and as a consequence of that, so did the incidence of illness. Perhaps it was the too rapid change-over or "deregulation" on the surface, which led to the disappearance of boundaries and tradition. The market's need for flexibility and rapid adjustment increased pressure on the democratisation process and the team approach where people had responsibility for their own decisions. Individuals were confronted with possibilities, expectations and the need to participate and exercise influence and this led to increasing stress. On the other hand, there was too little educational awareness and psychological insight about what such an extensive system change entails.

Sennet[231] holds that that we introduced teamwork, flexibility and more individual responsibility, but simultaneously threw aside tradition and authority:

Chapter 11

> The repudiation of authority and responsibility in the very superficialities of flexible teamwork structures everyday work life as well as moments of crisis like a strike or a downsizing (p. 114).

The net effect was certainly greater freedom, participation and greater influence, but without entailing greater control for the individual. This freedom was unjust because it still involved a significant element of subservience. The work situation involved more freedom, but primarily it became a freedom to work as much as possible, and indeed to work oneself to death. It is therefore important to reflect on why we wish to have participation, who gains from it and what conditions apply. Participation can be a goal which is good in itself and simultaneously, a means of attaining other values, both for the individual and the organisation.

The possibilities and conditions for increased participation can be discussed, planned and otherwise dealt with as a structural and organisational issue. Participation in itself is, however, more of a process than a state of affairs. Participation can exist on paper or be talked about, but it is individuals' attitudes and behaviour which ultimately give participation its concrete reality. It is in individuals that the feeling of "belonging to" and the wish to take part has to exist.

With such an individual perspective, we can assume that if people feel well their determination to participate increases. Participation is good for health and health is good for participation. This seems circular, but it can also be a picture of a good development spiral in a workplace.

Conditions governing participation

How is participation created?

This how-question is perhaps the most crucial in both the overall work of bringing about change, and in our reflections about health promotion as strategy. It is comparatively simple to make plans and mobilise practical resources. But how do we mobilise people? How do we deal with the intervention in the social system which makes the individual person a real participant? In other words, individuals are given a real influence over the daily agenda and are active in drawing up plans, putting across ideas and are generally involved in a balanced way.

Participation can neither be arranged to order, nor is it something which can be commanded. Participation takes place when circumstances which are favourable to it are created. We shall discuss three areas which play an important part in creating participation in an organisation:

- Learning
- Communication
- Leadership

Learning

According to Norrgren[232], in his description of two main strategies for bringing about change, both *learning* and concern for the human beings in the system are important prerequisites in working life. The first strategy he calls *programmatic change* which is characterised by careful planning, steering and control. Most things are thought out and agreed in advance, preferably in a *top-down* perspective without involving more people than are necessary.

The second strategy, Norrgren calls *learning-strategy*. This means that maximum attention is paid to the *lead-up* to the process of change itself. The participation on the part of every employee,

Chapter 11

dialogue and equal status, form the basic assumption in an analysis of the problem and suggestions about how to solve it. The employee is an actor and acquires an increasing understanding of where the company is headed. A learning process about change has begun. This process generates knowledge about joint answers to both the what- and how- questions.

The learning strategy is followed by questions concerning views about human beings, knowledge and education. In contexts where learning and participation are the goal, it can be profitable to strive after a view of human beings which is akin to the more pedagogical view proposed by Siv Their[233] (fig. 11.1)

See human beings to a lesser extent as:		See human beings to a greater extent as:
Objects	→	Subjects
Steered by genetic factors	→	Steered by environment and influence
"a well" (to be filled)	→	Spring
Receiver	→	Giver
Physical being	→	Mental/Spiritual being
Completely trained	→	In a process of learning
Limited	→	Unlimited
Static, Passive	→	Dynamic. Active

FIG. 11.1 *A more pedagogic view of human beings*

If such a value-shift becomes internalised among top management and the people in charge of the process of change, their attitudes and manner towards the employees as human beings will alter. In their personal encounter, this way of looking at things, or the picture they have of human beings around them, will become real: as someone has said, *what you see me as, I become.* This view of human beings also becomes the basis for a learning dialogue which concentrates on asking, listening and answering more than on saying what one wants oneself.

Our view of knowledge is linked to our view of human beings. When the individual's role and importance is to be strengthened both with respect to knowledge and organisation, the *quantitative* view of knowledge and educational technology must be largely replaced by a *qualitative* view. Knowledge is then treated as a perishable commodity which changes all the time and is renewed. As Maltén[234] puts it, real knowledge is a way of being, rather than a capacity to remember things and reproduce certain facts. Learning's task is to impart *structures of thought* which provide a *problem-solving capacity* and a *readiness and capability to act*. It is *active* as opposed to ready-made and pre-packed, or in the words of Maltén, *passive knowledge*. Obviously Maltén does not exclude fundamental knowledge and skills as an important foundation.

As we have noted earlier, health promotion is based on an actor perspective, where expert knowledge has a different content and role from that in the traditional programmatic work of bringing about change. The material, which the expert can supply, is assistance to reflection, the formulation of goals and problem-solving. With the help of an overarching theory and knowledge of processes, the consultant and person in charge of the process of change can contribute to what Bjerlöw[235] has called *meta-learning*. The workplace evolves its own shared knowledge of how things are to be solved and how the change is to be brought about. This form of learning is, just as we have described the change in general, more of a process than a set of predetermined plans and moves. It is knowledge which helps the people in the workplace to deal with several levels of abstraction, rather than with practice and action. A workgroup with this form of knowledge does not begin by proposing measures, but instead first of all reflects over the aim involved, the choice of method, the consequences and thereafter poses the question- what do we want to do?

How are we to look upon pedagogy in a learning strategy in the workplace? Given the view of knowledge, the view of pedagogy is that the starting point is human beings themselves and what occurs inside their heads. This focus is quite different from traditional view which regards teaching as a matter of conveying knowledge, a view which is still common both at school and in working life where the starting point is what goes on in the head of the teacher or consultant. "Knowledge" is then conveyed in the hope that the maximum of what is communicated is understandable, interesting and will lodge itself in the participant's head. This teaching approach is still in operation, despite the fact that research and practice have taught us that only a fraction of what is conveyed in this way has more than a momentary entertainment value.

> A learning strategy which aims at change, must apply a teaching method which helps people to question ingrained ideas. Docherty[236] et al. refer to Argyris who holds that:
>
>> True learning means cognitive, and even in some cases affective, development which is possible only when people are aware of and critical about their existing cognitive patterns, their "theory-in-use". Very often defences to reduce anxiety get in the way of these profound learning processes. Rather than transforming their thinking, people build fortresses around it. (p.10)

In order to achieve such genuine learning, the teaching method must adopt as its starting point the actual workplace's unique situation and the people who work there. Most important is the daily dialogue between the person in charge of change, the "idea-bearer", and the other members of the workforce. The daily dialogue can

Conditions governing participation

be reinforced with teaching opportunities. These ought to involve several senses and not just hearing and sight. An innovative educator like John Steinberg[237] has suggested that the instructor should plan for *structured flexibility* and create learning in an *organised chaos*. Organised chaos consists in creating situations, where people feel secure but which break with received ideas and try to encourage humour.

In all forms of teaching situation, the teacher or person in charge of bringing about change has the task of dealing with their own needs and values in a way which is objective and constructive and does not take away the initiative from the people who are affected by, and involved in the process. Moldaschl and Brödner[238] question whether this is possible. Reflecting about things and working in pairs with mutual feedback can be ways of dealing with this. It is about meaningfulness (fig.11.2)

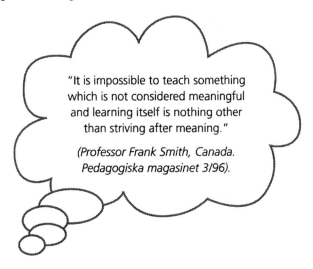

FIG. 11.2 *On meaningfulness and learning*

Chapter 11

Communication

Communication is another domain which has importance for participation. When people are to be made participants, ideas, knowledge and information must be processed and communicated. Increased participation also means that the continuity and intensity of relationships increases which leads to more friction and more conflicts. Lennéer-Axelsson[239] holds that in a workplace we can seldom freely choose our workmates, while simultaneously enquiries show that people prefer pleasant workmates like themselves. There is therefore greater demand for the toleration of human variations. Sometimes, this is rendered difficult, because people do not know one another or speak different languages.

Dialogue- conversation between people- is the key to participatory communication. It is seldom expressed or written down, but the absolute first step on the way to participation is that people learn to know each other as persons. Some form of arrangement or activity is needed which allows people to come closer to one another, and feel free to speak in a more personal way. It helps if people are released from established patterns and roles, preferably in another environment with elements of physical movement.

In the case of newly formed groups, some form of team-building or "getting-to-know-you" activities are often organised with the aim of creating a positive, friendly atmosphere. This should also be a recurring feature of established groups and especially when the composition of the group or the conditions it works under, are altered. When no time is given to this, the "getting-to-know- you" process takes longer. In the process, there is a danger that people's brains are more occupied with the person's role and position in the group, than with how colleagues view matters and above all how the task is to be solved.

Conditions governing participation

These are unconscious psychological mechanisms which are also linked to the maturity of the particular individual. A secure sense of maturity can be based on good self-knowledge and is expressed in a clearer *other-orientation*[240]. These phenomena are somewhat diffuse, but it is possible to make them visible and deal with them constructively. Thelander[241] has given an example of a hospital which sets out to create an organisation with *adult roles*. It was considered that in the traditional hierarchically structured organisation, the forms for management and communication are a barrier to participation. When information comes principally from one particular side, there is a great risk that people who are on the receiving end, adopt a subservient role and, as a result, surrender their responsible adult roles and adopt instead the more irresponsible role of a child. Reflection about this gave rise to a striving for more adult attitudes and ways of behaving, a different view of the manager's role, and a clarification of the decision-making arrangements which would reduce the risk of a flight from responsibility and a search for external solutions to the problem.

Communication and participation function with different degrees of success, depending on a chain of circumstances. It is a matter that involves individual people's maturity and skills. It is also affected by organisation, the location and premises and other physical circumstances. As we have already mentioned before, there is no tailor-made pedagogic goal for either the health promotion process in general, nor for the efforts to create participation. The conditions which apply to a smaller, team-based organisation are quite different from those for a large hierarchical organisation. When we are working to bring about greater participation, a workplace with a low level of education and language difficulties needs to be approached quite differently from the way we would approach a lawyer's office or a university.

Chapter 11

Any person, who is going to take part in an initiative which aims at participation and dialogue, must ask themselves a number of questions. How important is the size of the organisation? What sort of participation can be achieved when people work at different levels and in different places? How important is the physical design of the workplace? It has proved easier to create an elementary school based on teaching-teams and with good possibilities for everyday dialogue, if one simultaneously builds a new school where the premises fit in better with the work routines to be adopted. By providing a personal workplace, with nearness to colleagues and natural meeting places rather than long corridors with rows of classrooms, it is easier to promote participation both for pupils and teachers.

How does the nature of the work affect possibilities of creating participation? Is it possible to create time for systematic communication? The work is perhaps divided into three shifts with certain people - e.g. in a care establishment- only working nights. In another workplace, the workforce is spread out and is to be found sitting in their cars etc. Ordinary daily clinical work, emergency work, night work, distance work, global localisation- the list is long over factors which affect the possibility of communication and how we can experience participation.

All workplaces with at least two people, can be seen as a setting where there are -or arrangements can be made- for natural meeting places for a meaningful exchange of views.

Leadership

Leadership is the third important area for participation. When managers strive towards greater participation, it means delegating a great deal of control. A higher degree of participation in a workplace means a greater degree of chaos, a lesser degree of predictability

Conditions governing participation

and perhaps less possibility of steering things. To do this, we need managers who feel personally secure and have courage. To dare to rely on people, to dare to show that you rely on your workforce means that as a boss you create mutual trust. "As the boss, I show that I rely on you and when I do so, you are strengthened in your role and simultaneously look upon me as a wise boss." It is easier to have trust in such a manager.

Tomas Backström[242] holds that when the boss dares to rely on his employees and associates and on the often hard-to-grasp learning process in the organisation, organisational learning can take place which develops a competence which is shared collectively by the entire workforce. Participation is created by social processes in which communication and above all dialogue are important leadership tools. From this perspective, leadership, according to Backstrom, consists neither in steering or inspiring with enthusiasm, but rather in arranging and directing the preconditions which stimulate the social interaction which in turn promotes the learning process. This learning process is then a crucial precondition for creating a sustainable organisation, which Backstrom, Eijnatten and Kira[243] describe in the following way:

> A sustainable work system is, according to the complexity metaphor, most efficiently grounded in individuals acting and interacting without top-down or external control. Managers create and nurture conditions and affordances that allow alignment to emerge, but any such alignment must originate in the individual. Alignment is a matter of internal interaction and dialogue, promoting collective learning and self-organisation, which creates a common culture, a flow of information, a frame of interpretation and a common vision. (p.67)

Chapter 11

Being a manager and practising leadership which stimulates participation thus rests on liking and wanting to be with people. A competent manager must also have the determination, capability and physical and mental energy to communicate with others. In addition such a manager must possess a practical teaching capacity to be able to create the necessary preconditions for dialogue. Participation is an important precondition for creating participation and managers and leaders must have the determination to be inclusive. This part of managerial development is thus perhaps one of the most important areas to begin with in creating participation.

Theorell[244] raises the question of whether it is possible that increased participation can be a way of improving the state of health in Swedish working life. He holds that:

First of all, it is not really known how to bring about greater participation in workplaces. Some of the larger-scale experiments, for example, of " tearing down the pyramids" in the public sector, have probably been ineffective in the sense that they have not led to any greater participation on the part of the employees in any real sense.

One possible conclusion is that overarching organisational changes do not lead automatically to greater participation. The process of creating participation needs some measure of structure and planning which involves employees in a long-term process of change. Thus what is needed is an *educative approach* and a *pedagogic leadership* which can guide us on the way to change. Pedagogy has its starting point in human beings and is directed towards a goal. The path to that goal must have a conscious structure and contain elements which encourage dialogue and learning. One can read how this is done in teaching literature etc. dealing for example with organisational change.

Inclusive teaching strategies will in the future be more common in situations where health educators, process managers, researchers and others co-operate to run health promotion projects. In Sweden, the study circle is a traditional model for this. There is a counterpart in other countries where various types of "health circles"[245] are organised within companies. In other workplaces, arrangements are made for "health development days", analyses of the current health situation in the form of *work-shops, weekly meetings* with themes and *beehive discussions, focus groups, future workshops, dialogue walks* and *Future Search-conferences*. These are some examples of concepts which involve pedagogical methods where the participative component is central.

Aspects of participation

Participation and legitimacy

What confidence do the various actors have that organisational change and development work really can lead to better conditions in workplaces? Brulin and Ekstedt[246] maintain that the legitimacy of organisational development work in Sweden declined particularly during the 1990s. Hard restructuring in working life which preceded this period, helped not only to increase absence due to illness but also led to researchers and the media spreading a picture of desperation and hopelessness, which undermined the whole of working life's legitimacy. Many workplaces which in different ways had worked with organisational change began the 1990s more tired of change than they were filled with energy to develop.

The connection in Sweden, between the change in working life and the increasing incidence of absence due to illness is certainly difficult to interpret, but the negative media picture and the weariness with change which many had personally experienced, contributed probably to reduce confidence both in working life's actual qualities and its capacity to change for the better. According to Brulin and Ekstedt, the lack of legitimacy-creating activities organised by employers and trade union organisations during this period, also helped to diminish confidence in the utility of work dedicated to development and change.

Now reality is not simply black or white. There are organisations and individual workplaces which have had a positive development with regard to productivity and the health of their employees. The workforce has a belief in the future and looks upon its workplace as a place where change and development take place and are directed at important goals. Sometimes these good examples are publicised and can help to increase legitimacy. Bo Rothstein[247] holds that a positive change in a wider context presupposes that the parties involved take an active part in further processes to create legitimacy and trust. Legitimacy thus depends on the fact that several levels participate. From a branch and organisational perspective, for example, employers and trade union organisations contribute to a belief in the future and legitimacy. In every workplace, the workforce needs to feel that they have the management's support and the confidence of those around them.

Resistance to change

When the process of change is about to get under way or is under way, there are certain people who want to be part of it and help to ensure that the process will run on. There are also those who do not wish to help, but perhaps quite simply wish to resist the process of change and to make it more difficult. The type of reaction depends, both on the individual's capacity to accept change and the way in which the change is introduced. In a given workplace, there can be a complete spectrum of personalities from those who are enthusiastic about change to those with a great need of stability who become stressed simply by hearing the word "change".

In every situation, where the person leading the process of change has to operate, the preconditions can vary greatly. Resistance can sometimes be predicted, but it can also arise from an unexpected quarter, both from above and from the side. By being aware how normal scepticism, fear and safety in tradition steer people's attitudes to what is new, we can be somewhat better prepared and employ wise tactics.

Participation takes time

Ideas-based investment in development and health projects can be "hard to sell" and motivate, both to management and workforce. The links between the idea and aim of the project, the contents of the program and ultimately to those needs in individuals and the workplace which can be considered to be satisfied by it, easily becomes abstract. As a result, creating participation takes time and it must be allowed to take time. Preparatory work, the work of anchoring it in the organisation, information and discussions require to occur in stages in a way which makes the purpose comprehensible and helps people to make up their minds.

The part of the process dedicated to explaining and gaining support is a question of getting the message over and reaching people. The way this is done needs to be adapted to take into account the uniqueness of each situation. Detailed planning and the apportioning of the time available, are determined by the number of people, the organisational structure, culture and earlier experience of work dedicated to change. Just as there has to be time for this in the calendar, time must also be set aside from the actual work to communicate and have a dialogue, a cost which, according to research, repays itself later.

Chapter 11

From fitness and wellness programme to communication to organisational culture

When the idea is abstract and the goal has not been completely formulated, it can be a help to impose certain limits and to make things concrete. The long term aim can be to create a health-promoting workplace in a learning organisation which contributes to people's job satisfaction and health, as well as to the organisation's performance and long-term sustainability.

The first step can very well be a fitness and wellness project which promotes co-operation, job satisfaction, and the physical and mental energy needed to cope with everyday tasks. Such an investment is easier to "package" and sell when the specific program is closer to the individual and can be given concrete form at an early stage. When a process of this kind has finally been launched and news of its value has got around, one has a good basis for building further for the long term. With the help of what is concrete and down-to-earth, understanding and the wish to participate can grow more quickly and thereby allow one successively to tackle more abstract phenomena such as people's view of their profession, their attitudes and behaviour and the common culture. An account from the south of Sweden illustrates this.

The story of the health promoting workplace in Malmö

In a project which had been well-anchored, the conditions determining health in the workplace were to be strengthened. With the help of a health questionnaire and a workshop involving the whole workgroup, an analysis of the current situation was carried out. All in all, five important areas for initiatives were chosen together with the workgroup for the coming year's environmental and health work.

Work would be concentrated on:

- retaining the physical and mental capacities needed to cope
- breaks and rests
- decoupling from work ("distancing")
- making the workplace more attractive
- a more even distribution of the work between the people involved
- Activities relating to these areas were organised regularly in the course of a year.

Afterwards, there was a check-up to see what improvements had been noted by the personnel. It was found that there had certainly been improvements in the five chosen areas. However, there were also improvements such as greater comradeship, trust and feelings that one was respected.

There is possibly a danger of getting stuck at the concrete stage and failing to advance to the underlying human questions. Interest in continuing after the involvement and the enjoyment of the novelty of the fitness/wellness activities has declined, relies on the fact that there are the human skills and the necessary "tools" for dealing with the more abstract dimensions of the workplace in the continuing learning process.

Too much participation?

Participation has been a term of praise and an obvious condition when working life should be organised, led and developed. A great deal of research which extends from work on physiological changes linked to differing degrees of control, to sociological studies of how

Chapter 11

organisations and societies function, have shown the importance of people's participation. There is therefore today no cogent reason to question the importance of participation. But the world around us is full of nuances and general conclusions are always subject to certain reservations. For that reason, there is need to remind ourselves about the importance of thinking critically and reflecting about what we do.

Huzzard[248] points out the link which has been observed between increased participation, responsibility and possibilities of exerting influence on the one hand and increased ill-health on the other.

> Recent international statistics on stress, burnout and healthy work organisations suggest that many 'newer' organisations are consuming rather than regenerating their human resources (Docherty et. al, 2002a). There are indications, moreover, that many 'empowered' workers lose a sense of direction in their work as well as clear limitations on what they should do, how they should do it and during what timeframe. In the words of Docherty et al. (2002b:3) ' ...where bureaucratic structures and rules have disappeared, they have left the mature adult lost, lonely and increasingly stressed'. In other words, traditional sociotechnical design assumptions on autonomy and empowerment may need to be rethought in the light of our post-bureaucratic forms of organising. The possibility that empowerment and autonomy can actually increase stress and contribute to burnout is put into stark relief by the findings of the Third European Survey on Working Conditions published by the European Agency for Safety and Health at Work in 2000. Of the 15 EU countries covered by the survey, Sweden registered the highest percentage of workers- 79 per cent- whose pace of work was dependent on direct demands from people such as customers (as opposed to management). (p.61)

Conditions governing participation

This picture describes how new conditions and organisational forms have been introduced into working life. The idea had been that both the individual's situation at work and market utility would benefit from the diffusion of power and responsibility described above. Instead it led to greater stress since forms for decision-making, learning and problem-solving failed to be sufficiently developed and adapted in the majority of workplaces.

Health promotion aims at an integrated process of change which is focussed on health and is carried out with considerable organisational participation. This process can also be a stage in a necessary transformation which many modern workplaces and organisations must undergo. It is a matter of developing work forms involving dialogue and learning in a way which diminishes the destructive intensity of working life, and instead releases energy and creates sustainability and the preconditions for health. It must also be combined with activities which reinforce the co-operative efforts of the work group and the individual's physical and mental capacity to cope in daily work.

There is thus reason to reflect about what we mean by participation and why it is desirable in a particular concrete situation. Every workplace has presumably its own appropriate level of participation which allows it to balance the parts with the whole, the individual with the organisation, freedom with regulation and chaos with structure. . Much development work remains to be done in this direction.

Summary

Participation is central to workplace health promotion and in work to bring about development and change in general. By creating greater participation, we can achieve important effects such as greater understanding, greater involvement and a better adaptation

Chapter 11

of the planned program of change to the particular workplace and organisation in question.

Participation is often much discussed and well anchored in many situations. There is everything from legislation to pedagogic models of change which underline its importance. Different people assign different meanings to participation and differ in their expectations of what it can achieve. We have examined a field which describes participation as a question of power and influence (empowerment). Another dimension pertains to how human beings balance being independent on the one hand, and wishing to belong on the other. A further aspect of participation is linked to democracy and justice which are deeply rooted in Swedish working life.

The third question asks how participation is created? To what extent is it a matter of learning and knowledge? What role is played by communication between people and the purely practical possibility of communicating? We have also touched on the management's importance for participation. The educative side of management has presumably great importance in getting the personnel to participate constructively.

There are also other reflections about what participation means. Participation occurs in human interaction. Influencing this takes time and can encounter opposition. Both practical wisdom and a well balanced code of values, which allow management to deal both with organisational goals and people's needs, are required. Failure can lead both to bankruptcy and burned out people.

12 Process orientation

In this chapter we discuss the fourth criterion which sets out to try and answer the question: how do we create a movement dedicated to improving health?

Hitherto we have described a focus on health promotion which in everyday language means making use of health determinants. This is the basic idea and approach in order to work with the conditions for health. To do this we need knowledge about what is good for health.

Secondly, we select the workplace as the setting and starting point. It is in this particular context and with the people who are to be found there, that the work for better conditions for health is to be carried out. Organisations are systems with general characteristics which must be taken into account. Moreover, every particular organisation and workplace requires us to adapt to its own particular situation. People's lives outside the workplace also mean a great deal for their health, but can only indirectly be affected by the organisation's health work. The hope must be that health and environmental work within the workplace will be sufficiently successful for the setting to become a resource for life in general.

Creating participation is the third criterion which must be satisfied if interest, collective understanding and behaviour are to be influenced. It is a matter of the work group's own organisational and social situation and it is these people who have to act. Those who work in a particular department are those who know best how things are there, and what needs to be done.

As a logical or necessary piece of the puzzle, may be added the process-oriented approach. Because of the complexity in an

Chapter 12

organisational /social system and the unpredictability associated with change, an understanding of and a capacity for applying a process-oriented approach, is the only possible way to bring about and steer a movement towards better conditions for health in this setting. In other words, we accept the view that the work of change and development in organisations cannot be programmed in advance.

Why process?

A working life, which is moving towards greater diversity and less central control, needs to replace, bit by bit, static and programmatic conceptual models with dynamic and process-oriented forms of development. Health promotion, as a strategy for change, which aims at creating the requisite conditions for the health of individuals and company performance, must submit to these principles. Cressey and Docherty[249] underline this as a requirement for organisational performance and sustainability:

> In our current turbulent and changing work organisation environment a one-dimensional approach is insufficient and possibly damaging. The key issues for organisational performance in both manufacturing and service industries are diverging from simple Taylorist strategies. The dominant and active elements in that paradigm had to do with control, fragmentation, routinization, stability and replicability of performance. Contrast this with the core issues now dominating work organisational discussions- creativity, commitment, reflexivity, learning, chaos, quality and dialogue. (p. 175)

Today, several organisational theorists hold that an organisation which consists of people can be best understood by seeing it as a living system. The organisation becomes an open system which is continually exchanging information and energy with the world about it. According to Ahrenfelt[250] such a system has an inbuilt capacity for self-organisation which enables it to adjust and survive changes taking place in the surrounding world. From this point of view, successful work in bringing about change implies that employees can be activated and contribute with their energy and creativity.

The processes in organisations and markets, as local and global phenomena, are also complex and difficult to understand. Sometimes we cannot even guess what causes what in an organisation or market. Ahrenfelt uses chaos theory as an example to illustrate this complexity and unpredictability. He holds, however, that even if we have an experience of chaos, it does not need to be *wholly chaotic*: it is rather a matter of a relative experience of chaos. A person with a knowledge of the whole and knows how the system works, does not consider that it is chaos. The purpose of making this comparison with chaos theory, is to make people understand that we gradually leave a situation of stability and predictability and are instead forced to deal with complexity, uncertainty and change. Uncertainty and the feeling of chaos can be decreased by an appreciation of the holistic approach – keeping in mind "the big picture"- and knowledge of the processes which take place in living systems. According to Ahrenfelt, organisations should be seen as living systems which operate in a living environment.

An understanding of "the big picture" and the process in the unique situation becomes, according to the foregoing way of looking at things, a necessary starting point for the work of change. There are no ready-made solutions for creating the health promoting workplace which can be determined in advance or which fit every case. Of course, there is general knowledge about what constitutes effective work, good environment and a wise life style. The initial assumption, however, is that every situation has its

Chapter 12

unique preconditions and if the health promotion process is to yield results and be experienced by people as something meaningful, every workplace at the local level must be in charge of shaping the process.

What is a process?

A process can be seen as a (protracted) sequence of events which entails that something undergoes change or development. In other words, a temporal dimension is involved with the process having a beginning somewhere and (perhaps) an end at a later point of time. Even if the process does not have a final goal in time, the majority of processes which people create are designed to accomplish something.

The process can lead to *development* or *change*. These common concepts ought to be defined and distinguished by those working with processes. If the purpose is development, the process is directed at a goal or a better state of some kind. Sometimes however, the demands or needs are so great that a development of the present ways of working does not suffice to reach that state of improvement which constitutes the goal. The demands placed on the enterprise have become so different that the current organisation is completely outdated. Perhaps several development projects have been undertaken to solve the crisis or deal with the problem without success. The only way of making some advance is then through something more radical,, namely a change or replacement of the system. In such cases, we can, according to *theory of logical types*[251], speak of a change of *the second order*. The previous attempts to find a solution which were directed at *developing* the existing enterprise can be seen as a solution of *the first order*, that is to say, a solution within the system or by means of those methods of solution that we customarily employ.

When we use the concept of *change* for a process, this indicates that we must go outside the present system and tackle the conditions of the system by working on the surrounding system levels. In everyday terms, we would say that there is need for a radically different solution which may require considerable creativity to find.

Apart from its purpose of accomplishing something, the process has a character which cannot fully be described. It can be seen as a sequence of events. It is interesting to reflect about what this sequence of events consists of. Is it possible to describe it and to what extent is it possible to influence it?

Technical versus human related-processes

A person speaking about processes or trying to bring about change where processes are involved, ought to be aware that the word 'process' conjures up different images in people's heads, depending on whether one is speaking to engineers, biologists or behavioural scientists. It is absolutely necessary both for the process manager and the rest of the team to be clear about this, since the meaning attached to the concept can represent completely different and opposite values and ways of looking at things.

Traditionally the most common use of the concept of process is in technology, chemistry and other domains of natural science. In such domains, process is a predetermined sequence of events where every stage in the process can be identified and described. It can be a chemical process such as the manufacture of liquids with certain properties. It can be a process in the paper or automotive industries or in some other kind of manufacture. Marmgren[252] calls such technical processes, *hard processes* and they are characterised by the positivistic predictability of natural science. The events making up such types of process form causally determined linear chains. According to Marmgren, often such hard processes focus on speed, efficiency and rationalisation.

There are also *human-related processes* (or so-called *soft processes* to use Marmgren's terminology.) Here one means all the hard-to-grasp, complex and sometimes irrational phenomena which affect people in an organisation. Describing and trying to handle such processes is based on an entirely different type of knowledge and takes place in completely different circumstances. Ahrenfelt[253] has a definition of the concept of process which clearly distinguishes the human aspects from the other parts of the organisation:

> The concept of process is the combined interaction in the field ,"the result creating process", between the cognitive and emotive structures and patterns of thought which are loaded, activated and in motion within the system and between the system and context. (p. 227)

It is thus about all the thoughts, feelings and actions which take place within the individual, group and organisation at large during a given period. This process takes place all the time and must have the highest priority, since it is already in operation before we enter the picture and start process-directed work aimed at development or change.

Concepts associated with technical processes such as speed, precision, quantification and control, do not fit in here. Instead, such a way of thinking becomes an obstacle in dealing with human-related processes and simply causes damage. Human processes must be dealt with on the basis of human conditions, through dialogue, learning, understanding and things must be allowed to take time. This would seem to lead to a major confrontation, when the rationalisation demands of working life are hard.

Process orientation

The processes of an organisation are seldom purely technical or human-related. People work together in and with the production system and the organisation imposes conditions which make the human-related processes subject to time, performance goals and other hard data.

Health promotion as a strategy presupposes an appreciation of human-related processes, but also a capacity to give the process a structure and systematic form which successfully integrates the human-related aspects with the technical processes in workplaces. The task for the process manager is both to bring about a better understanding of those human-related processes which create results and to counteract a tendency for them to become what Ahrenfeldt warns us about- namely a basically static model described in process-type terms.

A change towards a more conscious treatment of human-related processes takes time for many reasons. Both the organisations which buy health services and the consultants who deliver them have to build up this form of competence. It will therefore probably take a long time before the majority of workplaces use human-related processes to bring about human-related changes. In the longer term, the goal is perhaps to achieve a culture of change where the technical and human-related processes work together in development and learning.

In what follows, what we call processes refers mainly to human-related processes, that is to say the social interactions in organisational systems which do not allow themselves to be planned in detail and to be steered in the same way as their technical counterparts. In order to clarify the use of the concept of process in health promotion, we may distinguish four different "levels" of human-related processes, which we review in the figure beneath and in the following explanations:

Chapter 12

> **Process Level 4**
> A step by step social change aiming at better possibilities for people to influence their health.

> **Process Level 3**
> A human-related process which is conscious, creative and contributes to sustainable change and learning.

> **Process Level 2**
> A human-related process which makes use of a "hard" structure in order to be clear and to get going.

> **Process Level 1**
> Is essentially a technical process but with a "human-related" content.

FIG. 12.1 *Levels of the "soft process orientation" in work dedicated to bringing about change*

Process Level 1

The first process level, which we describe, ought really not to be included here since it deals with technical process's conditions but with the content of a human-related process.

The procedure is in this case based on the idea that an expert or manager knows what's best and therefore can and ought to determine what's best for a workplace which is included in the process of change. The management wants to carry out the work to bring about change quickly and efficiently according to certain predetermined goals and a predetermined plan. An analysis can very well be included in the preliminary work but takes the form in practice of collecting data which the expert then interprets and uses as a basis for the measures he subsequently proposes. Such a programmatic attitude and response has been shown to be utterly ineffective since this way of bringing about change seldom succeeds in involving the people in the workplace at a more profound level. The program to bring about change becomes instead a "theatrical piece" which the management hopes is happening in reality. However any notion that the change really occurs on the workshop floor and in the individual employee's understanding, attitudes and behaviour, is a myth in which only the uninformed believe.

This approach is also unfortunately practised under the heading of health project, where a consultant first finds out about the life style and well-being of the workforce and afterwards suggests measures intended to make them feel better. This procedure assumes that the capable consultant/ expert know exactly what the person who is overweight, in a bad physical condition, smokes, or is unhappy in their work, has to do. Ironically one might say that the only thing that is lacking is that the consultant does not know if the employee in question -"the health object"- wants to be "subjected to" or personally undertake the measures proposed.

When the consultant finally perceives that people's knowledge, values, opinions and behaviour are so-to-speak "human software", he/she perceives that they also must be understood in terms of human-related processes.

Process Level 2

Working life is traditionally conceived in terms of technical (or 'hard') processes and health promotion and other human-related process work has to be integrated with this tradition. It is therefore reasonable to assume that it is usually a technical process which conceptually forms the starting point when process work is discussed in the organisation. The understanding and knowledge possessed by the individual organisation varies and influences how far it becomes involved in thinking about human-related processes. When the latter way of thinking is to be introduced, it should be done, step by step, and should preferably begin by taking the description of the technical or 'hard' process as a conceptual model.

A first step on the way to human-related processes of change may be to make the hard process human-related or describe the human-related process by means of hard concepts so that people can recognise what is involved. For example, this can imply that the process has a clear and "known" structure in the form of a" flow chart" with different stages. This process model is then implemented in a human-related way. In other words, analysis of the current situation, the formulations of goals, deciding what it is to contain, specifying the time allowed and checking-up take place afterwards in the process and in dialogue at the workplace. This approach implies that we have abandoned the traditional expert way of thinking where the consultant analyses what is needed and then puts forward proposals for what should be done.

Health promotion as a process to bring about change is applied in accordance with this approach at more and more workplaces, as a conscious step in jointly creating greater participation, better conditions for the health of employees and improving the performance of the enterprise. We shall return to the description of this approach under the heading *structured human-related process* on page 308.

Process Level 3

The third level of process thinking can be seen as a development of level 2. It occurs when the organisation no longer requires supporting "hard" structures in order to bring about effective and sustainable human-related processes. The development of the conditions for health and the enterprise's performance are maintained continuously in a dynamic and conscious process.

Process orientation

The organisation or workplace which has attained this state has not achieved a level or quality which one can remain satisfied with. According to researchers, the health promoting workplace as a sustainable and regenerative system can never be a finished model, but is rather a continuous process. There is a dynamic in the organisation, where people have a feeling for the 'whole picture'; one finds openness, dialogue and other positive preconditions for collective learning which characterises this state.

Schein[254] maintains that senior management and employees, who have attained this level, are aware that the human-related process has great importance for the ability to make decisions and for working efficiency. According to Schein, a continual monitoring of the state of organisational interaction and communication is the basis for also maintaining favourable and effective human-related processes.

> Processes in the organisation follow patterns that can be studied and understood, and that have important consequences for organisational performance. Most important, processes can be rationally changed and adapted to increase the effectiveness of performance. (p. 178).

Backström, Eijnatten and Kira[255] hold that the human-related process is crucial in deciding if an organisation will be able to create sustainable work systems which contribute to people's health. According to them, sustainability is thus not a static state, but rather a process over time:

> Only a system that is continuously in a state of "becoming" can be called "sustainable". Sustainability cannot be regarded as a static characteristic of a structure or a process because everything in the system is constantly "on the move". A definition of sustainability

must take account of time as a key factor, and should focus on dynamic qualities of the system. (p. 65).

These are theoretical descriptions of how one can look upon the conditions for creating sustainability in an organisation and in a workplace. Some people may experience this as unduly restless and indeed a little hopeless. Should we not be able to work and be finished with our health promotion work? Surely life in a workplace is not quite so fluid and changeable as these descriptions would seem to indicate?

The complexity and degree of change differs of course from workplace to workplace exactly as other preconditions can vary greatly. Every organisation and workplace must draw its own conclusions about the conditions which apply. What may be generally valid, is the assertion that our society and working life are involved in change and this leads to a demand for adaptation. In order to cope with this satisfactorily, every individual needs to participate in the process: it is not enough to steer everything in the organisation from the top down. The form of level 3 process we are trying to describe entails that more and more people in the workplace become aware, get involved and want to help to create results which benefit both health and organisational performance. Ideally, in order for the work to function optimally, the workplace should become a culture of learning and collective responsibility. The positive energy for this is created by good relations, a sense of well-being, constructive dialogues and -to put it another way -a high level of the qualities which contribute to a *sense of coherence*: comprehensibility, manageability and meaningfulness.

This description can serve as a vision for what an ideal human-related process can imply. This vision does not contradict the fact that in everyday circumstances we will continue to structure, deal with and give priority to different questions at different points of time. We determine the agenda and carry out initiatives and projects aimed at development and improvement on the basis of need. The composition of the activities provides the answer to the *what*-question: what needs to be done or what do we believe are the most important things to do? The human-related process will be the answer to the question about *how* it is done and will determine the involvement, quality and result that is created.

Process Level 4

The fourth level of human-related processes lies at the periphery when we focus on the workplace as setting. We can describe it as an overarching process at social level, where the aim is to influence the overall preconditions for people's health, but according to the values and principles which are embodied in the words of the Ottawa Charter[256] relating to Health Promotion:

> Health promotion is the *process* of enabling people to increase control over, and to improve, their health (p.5) (present author's italics).

This process is subject to political control and its goals differ somewhat from the process of health promotion which takes place in a company or any organisation in general. At the social level, there is greater emphasis on striving for an equitable distribution of health. *Every human being should be able to satisfy their fundamental biological, social and economic needs.* This perspective is set out clearly, for example, in *The Foundations Theory* which Seedhouse[257] has put forward, where it is every person's equal access to the fundamental prerequisites of health which ought to guide the work of health promotion. This is a process which is influenced by society's structural, cultural and political preconditions. Both the

Chapter 12

way of describing and understanding this process, and the ways to influence it, differ from how this can be accomplished in an organisation.

The economic/material motives are used in a similar way and assume that if human beings have good health and feel well, their output will correspondingly improve in both quantity and quality, which benefits both society and the organisation. The organisation, however, has in general a more clear-cut goal: it aims in its activity to achieve a certain result, which strengthens its motive to promote the preconditions for health. A society does not disintegrate so rapidly as a company if some of its inhabitants feel bad or if the distribution of health is inequitable.

The organisation and workplace have an advantage in that they are limited and structured in a way which favours the establishment and direction of health promotion processes.

The structured human-related process

As we have mentioned above, there is no contradiction between thinking in terms of human-related processes on the one hand and working on the basis of clear goals with a structured work model on the other. Once again, it is a question of first *what* and then *how*. In this section, we describe a general model for a process of bringing about change. It is a "hard" model which sets out to ensure that important steps in the process of change are not lost sight of and that the activities which are needed to promote the human-related process are dealt with in a tactical and sensible way. It requires skills, resources and good planning when human-related processes are to be improved and when human-related processes are to help bring about a better workplace and improved work. The human-related process can in general be seen as a temporal dimension, a continual flow of change and learning in the organisation. At every

Process orientation

moment in time, there is a historical perspective, a situation as of now and a future perspective, but there is no definitive starting point or end.

The motive for describing processes as structured and human-related originates in the need both to make the process manageable and effective while at the same time giving the human-related part of the process its due. Structure implies that the process starts at a certain point of time and comes to an end in attaining a certain goal at another point of time. The start marks a definite 'now' or *current state* and the goal of the process can be described as a more or less hard-formulated *desired state*. The current state and the envisioned, desired state can be separated by a number of months or several years of process directed development work. It is worthwhile noting that when we describe a "current state" there is always a third temporal dimension of history and experience involved, because the past forms an important part of what has become "now".

The introductory description of the current state and the formulation of the desired state are necessary in order to be able to select important areas for initiatives, set goals for these and set priorities as regards the use of resources in general. In order for the ongoing process to be effective, there is a need for continual reflection and dialogue. We should continually be asking ourselves the following three questions:

- What are we to achieve?
- How are things at the present moment?
- What should we do next in order to advance further in the process?

The beginning and end of the structured process can in its simplest form be illustrated with the help of two "states" between which a shift or movement is then to take place:

Chapter 12

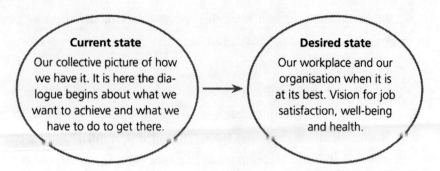

FIG. 12.2 *The structured process's beginning, direction and goal*

Fig. 12.2 can be a conceptual point if view which shows that the structured process has a beginning point and a direction. In the next stage, we can reflect about what happens during the process and what it ought to involve. Once more, it is the preconditions in every specific situation which decide the form the process is going to take. An example where this form of process thinking has been applied is the internal development project SALUT, which began at the University College of Trollhättan-Uddevalla in 2003 and ended in 2005. Using health promotion as a guiding idea, the aim was to develop a long-term process for improved health in the workplace. There was strong emphasis on participation and involvement at every workplace. Different activities intended to encourage and support the local initiative and bring about a local process were given priority during the time of the project. The success of the SALUT- project can be described in two stages. First of all, the project was designed to bring about improvements in the specific areas selected for the initiatives. Moreover, after the end of the project, the aim was that there would be a well-functioning system of environmental and health work, with personnel participating in a long-term process.

SALUT was run as a structured process in such a way that the project approach was used as a structure and organisational form in order to keep track and manage the process at a central level.

Process orientation

Moreover, in the implementation there was a clear direction to the process with process managers and a process group which constantly monitored what was happening and correspondingly adapted the content of the process. The three year project was summed up in three phases with differing main goals:

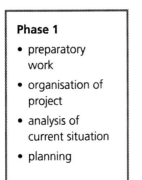

Phase 1
- preparatory work
- organisation of project
- analysis of current situation
- planning

Jan 03 – Dec 03

Phase 2
Local process
- initiate
- carry out
- develop

Central project
- support
- co-ordinate
- follow up
- develop

Jan 04 – June 05

Phase 3
- evaluate
- describe the form and content for the continuation of program for increased health in the work place
- how to go further

July 05 – Dec 05

FIG. 12.3 *The SALUT-project was divided into three phases*

Fig. 12.3 shows, on the one hand, the content of different parts of the project and on the other hand, the time which the different parts were estimated to take. In reality, the time limits were fluid in the sense that the phases intertwined with one another. Roughly speaking, during phase 1, the organisation devoted half a year to preparatory work, followed by a half year of analysis and planning. It is first in phase 2 that concrete things were carried out in individual workplaces. Finally phase 3 is not so often given priority in processes to bring about change. However, it is at this stage that the knowledge which has been generated during the project has to be collected and evaluated in order to serve as the basis for planning the period after the project i.e. the time when the local process is expected to run on its own wheels and when central resources and structural support are not available to the same extent.

Chapter 12

The project form can both be seen as a concentration of forces, a structure and a symbolic emphasis on the organisation's efforts to change and improve its work systems in the direction of greater sustainability and improved health. At the same time, the project is only a form and resource which has the task of initiating and facilitating the process in the organisation, thus helping it to be structured and on target, both as a local process in each workplace and as central process which applies throughout the whole organisation. To bring about a successful change, it is crucial that the local process is based on the participation and involvement of every employee. The central process has the task of dealing with the organisation's overarching interests and of creating conditions for change.

From process management in the SALUT project we learned that structured human-related processes of change can be described as a balance between extremes where only the particular setting and situation determines where the balance between the poles is to lie. See fig. 12.4.

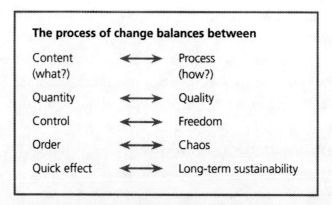

FIG. 12.4 *Opposing poles in the process of change*

The larger the organisation, the more difficult it becomes, to communicate and formulate the collective picture and get people to appreciate the point of the project. Why is this being done? What does it involve? How is it to be implemented? The following are some examples of such questions in the case of workplace health promotion:

- *What is the aim?* Put simply, we can say that the aim is to improve health in the workplace. This can, however, involve several parts which it is important to make plain. It can be about improving people's present state of health and/or improving the conditions for health. Another aim and perhaps the most important is to develop the forms and content in the workplace's continual and long-term work for improved health.

- *Is there some idea behind it?* In salutogenic health promotion, the conceptual starting point is to focus on promoting conditions for health.

- *Is there a more concrete way of picturing the goal?* The goal of the health work can be described at different levels. It can be to sink the level of absence due to illness or to increase "long term attendance thanks to health" in the organisation. In the specific workplace, the goal can be to create a balance between the demands of the work and available resources. At an individual level, the aim is to feel well, but it can also be about creating preconditions so that people are both physically able and willing to do a good job.

- *What is the role of the project?* A project is normally a limited initiative to solve a given problem in a given time, given certain resources. When the problem is solved, the project ends. In health promotion, the project form can entail a temporary investment of effort with the objective of finding forms for the health promotion work and

Chapter 12

developing them, above all supporting the development of processes at the local level.

How can we instructively describe the structured process in health promotion? It is really an unnecessary or impossible task, since every process is unique. Structure, form and content are decided on the basis of how the preconditions vary from organisation to organisation. However, there is a basic model in both project and process directed work which can serve as an example of the constituent parts and their order, which are needed in order to clarify the current state and create the involvement and movement towards the desired state. A process of change is based on movement and always contains several parallel processes. A complete account of these can therefore turn out to be very complex. The figure below shows various items which can be part of an overarching structured process over a longer time, or a subprocess which deals with an aspect or a special area of initiative in the work of bring about change. The headings we choose are something of a matter of taste and there are also variations when these types of process model are described in the literature. Fig. 12.5 is intended simultaneously to give the impression of development and movement along a time axis.

Process orientation

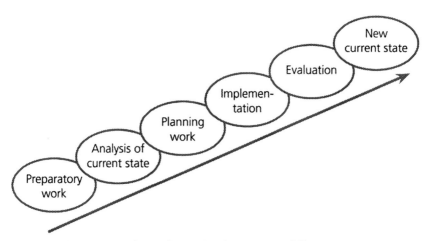

FIG. 12.5 *The structured transformational process at different stages*

The project form of work is appropriate and indeed can be necessary in order to create the preconditions and to shape and initiate a long-term process. It gives the organisational preconditions and structure which the process of change itself requires, to be effective. The preparatory work is extremely important for the whole sequence of events: it takes time and it must be permitted to take time. Consider the example of SALUT. This involved an organisation with around 500 employees, divided into some 30 workgroups. From the time the idea was hatched and the planning work was begun, a year passed before the actual project started. Afterwards, the first six months of project time was used for organisation and staffing of the project, followed by conversations with the senior management group, explaining and firmly establishing the project within the organisation, and other preparations. Only after this, did the project's analysis and planning phase get under way with a collective web-based health questionnaire and subsequent workshops in all the workgroups. This part involved analysis of the current situation, on the basis of which the workgroups then formulated the desired states which they personally wished to achieve. This allowed them then to also plan what to do in order to advance further with this aim. Parallel with this, planning took place in the central process group and in project groups, in order to spur on the activities throughout the organisation as a whole and to support and check up what was happening at the local level.

If the preparatory work is done well, this makes it easier to maintain subsequent work on a long-term basis, while at the same time, there is a greater chance that the long-term process will continue after the end of the project and can help the organisation to develop further towards more visionary goals. A visionary goal for the future can be to become increasingly a learning and participatory organisation with optimal preconditions for employees to feel well. In this way, it can also become a high quality organisation which also performs successfully.

Process management

In order to control and co-ordinate the ordinary work in an organisation, management and leadership are needed at different levels of responsibility and function. The type of business or activities conducted and the size of the organisation influence how appropriate managerial functions are designed, and also the relationship between technical and human-related processes. In modern organisations, initiative and responsibility has been shifted ever further from the centre of the organisation to its periphery and delegated to the individual person. Working effectively is much less the fruit of central management and control. It is much more a matter of individual employees understanding and assuming responsibility for the work as a whole and wishing to exert themselves. This principle also holds true of the structured process in organisational work dedicated to development and change, such as health promotion. Change is directed at, and dependent upon, the various decision-making and functional levels of the organisation. Each level must then be given the opportunity to assume responsibility and take part. Nor is it possible to manage and control the structured process of change from only a central level. The energy firing the process comes from people at various levels in the organisation who must feel themselves responsible for the overall picture. Boards, management and employees need to actively assume responsibility.

It is impossible to steer human beings- it is a matter of understanding and steering the process. The structured human-related process is steered by goals and priorities as well as by the roles, responsibilities and rules of the game which people in the organisation draw up. What then steers people's determination to participate are values and their appreciation of the aim and the situation. We all ask ourselves whether consciously or unconsciously: is this good? What's in it for me? People, who can see no merit in a project, are not going to exert themselves either.

Sandberg and Targama[258] maintain that people have different initial preconditions and interpret the same phenomenon in different ways. It is therefore necessary *to start from individuals' current understanding* (p. 159) when a new idea is being advocated. It is neither possible to describe human-related processes or to convince people of the value of abstract goals with the help of leaflets or elegant presentations. What is required is an educative process which is rooted in people's own ideas and imagination and where their reflections and their own words can be part of things, helping to formulate and explain what it is about. Only when this understanding exists and is felt to be important, can there be a commitment with regard to what is to follow.

Marmgren[259] describes the conditions confronting leadership in a human-related process:

- There is an objective or goal but it changes and becomes clearer during the process
- It is unclear what the key activities are and who is going to carry them out.

Chapter 12

- Learning gives new knowledge which provides new preconditions for future work.
- The process is creative and the plan is designed afterwards on the basis of what emerges.
- Management by positive feedback- amplifying what is right (in contrast to 'hard' processes which are steered by negative feedback - error-correction.)

The management of human-related 'soft' processes thus deals with putting up with uncertainty and daring to let things happen. There has to be a basic structure to follow, but there is no exact plan. Instead decisions are made on the basis of the interpretation placed on information which flows in all the time. In human-related processes, management is more about understanding what happens in reality than in being able to follow a pre-set plan. Patience is a virtue for the boss who feels that it takes too long to reach a result.

The human skills needed to manage structured human-related processes may be available in the organisation or can be provided by an external source. Schein[260] describes the role of process manager as being more modest than the traditional consultant-expert role. Participation and trust need to be established in a way which gives access to the human-related processes of the organisation. The process manager's role is not to propose solutions to the organisation's problems but rather to be a support in explaining the resources at the organisation's disposal and to find ways of using them in the process. According to Schein, the process manager can also help to make the organisation aware of its processes and their consequences.

Leadership in the case of structured human-related processes might be compared to the role of a theatrical director. The preparatory work is devoted to explaining roles and the plot or situation. Working together, the director and the actors interpret and put forward the aim of the performance, its underlying ideas and the way it is to be done. The actors have all the necessary skills and learn their roles. When the preparations are complete and the play is to be performed, the performance begins and each actor interprets and plays his or her role in accordance with the play's message and the work as a whole. The director is still a leader, but does not participate directly in what happens on the stage. On the other hand, he or she follows the performance closely and with insight, ready between acts or when the curtain has fallen, to initiate a dialogue about the play that has been performed.

Subprocesses

Processes of change in organisations involve several subprocesses and numerous activities in different areas. There are always and especially clearly in organisations with more than one workplace, both a local and a central process. The *local process* is really the main process that we are striving for. It contains everything which occurs in the different parts of the organisation. It involves the activity of each workplace, including staffs, service functions, even management groups, with their collection of individuals who have to function and communicate well in carrying out its task. The human-related process takes place among people at all levels and in all organisational situations.

The central process in health promotion work dedicated to change involves the functions and activities which are created in order for a local process to get under way, become integrated in the rest of the work, receive support and is followed up. This process arises from a central project group, process group, steering group or a particular process manager. For natural reasons, the larger organisation has a greater need, but also is better equipped to deal with this type of question. An important part in the central process

Chapter 12

is the follow-up and evaluation work which must be done in order to "steer", co-ordinate, use and spread ideas, knowledge and other results of what occurs in both the central and local process.

The structured human-related process can be seen as an overarching process in the organisation. It can also, as above, be divided into a central and a local process. We can further concretise and describe what the structured process can contain or consist of, at the respective central or local level. Here are some examples:

- Initiative areas
- Subsidiary projects
- Pilot projects
- Supporting processes

Initiative areas which require special treatment can be defined on the basis of known needs, or from the information emerging from an analysis of the current situation. There are perhaps gaps or needs in areas such as the development of competence, the physical working environment, worktime models, policy documents, information routines or personnel activities intended to create a pleasant atmosphere, which can receive special attention in a structured process. If it is decided collectively to give priority to some particular urgent area, it can then be appropriate to run this as both a local and a central process. The local activity represents the creative part of the process while the central activity creates resources, facilitates, co-ordinates and creates the resulting outcome as a whole. Certain initiative areas are by their nature more of central character, for example the revision of policy documents. Other areas which involve workgroups and individuals are much more local in nature.

Subsidiary projects can coincide with what are important initiative areas. What is suitable for dealing with in project form

Process orientation

can be organised as such and become an activity in its own right within the process as a whole. Such subprojects ought to be clearly defined functions or areas which concern the whole or major parts of the organisation. Examples of this are the following: devising a training course for senior management in health promoting leadership; investigating the possibility of introducing a working schema which arranges working hours in a way that is better adapted to the needs of individuals; creating a collective organisation for workplace wellness and recreational activities; designing a room in the workplace for stillness and meditation. When a participative process of change gets under way, smaller scale projects arise in every workplace. This is a sign that a change towards learning and development has started.

A *pilot project* can be an excellent way of developing and testing new work routines. The human-related processes of change are, however, difficult to test in an organisation as a whole. Either one has decided to initiate a movement towards change or one has not. It can be compared with wild water rafting where there is no point of return after one has "cast away" and there is no way of testing the strength of the current unless one has done so.

On the other hand, it can be a good way of learning and developing a method to run a structured process of health promotion in a part of an organisation. A section or local unit can "go first". It can be chosen on the basis of need or because it possesses sufficient organisational maturity to cope with a structured human-related process.

Finally, *support processes* cover the running activities or routines which are there to facilitate and assist the structured human-related process. An important area is communication and information processing. Education, documentation and economics are other

Chapter 12

support processes. As mentioned above, the important follow-up and evaluation process belong here.

Following up and evaluation of processes

Follow-up and evaluation are types of measure which can consist of everything from simple descriptions of what has been carried out to very complex analyses of the effects which a structured process has on an organisation. Here we shall briefly comment first on the concept of follow-up and then on the concept of evaluation.

Following up what is currently going on ought to be something which is done all the time in order to have up-to-date knowledge, to be able to judge the current situation and to make decisions. It is designed to keep the enterprise focused on its goals where the up-to-date information and descriptions decide if the process seems to be running in accordance with the underlying vision and goals, or whether some corrective intervention is needed. The collection of this information can take place in corridor encounters ("management by walking around") at workplace meetings or more formally by planned interviews or conversations. There are also other forms of collecting data, but the important thing is that the person or person directing the process is able to have their picture of the current situation updated and can act accordingly. People, who know their organisation, know in general the persons who are usually aware of the mood and views of the workforce.

In a structured process which runs over a longer time this updating ought to be integrated with the running process and should take place at regular intervals. If one makes use of technology, intranet or email can be employed at a reasonable cost in terms of work to ask some questions and obtain a good informational basis for decision.

Process orientation

Evaluation and follow-up can amount to the same thing from the viewpoint of method. In practice however, evaluation is a more systematic activity where investigations are made to evaluate the process or outcome of some initiative, program etc. which has been carried out. Evaluation which is done during the period when the process is going on -so-called *formative evaluation*- with the aim of adjusting or improving the continued work, can be placed on a par with the above follow-up work. The most common application of evaluation, however, is so-called *summative evaluation* which means that one waits till after the process has been carried out, to investigate and evaluate to what extent the goal has been achieved, what the effect has been in relation to the state of things at the beginning or other results which have obtained with other measures.

The evaluation of structured processes in organisations can be made into very complicated phenomena. There are in general a multiplicity of activities, professions and informal systems which both influence the outcome of what is to be evaluated and the reliability of the evaluation work itself. The complexity entails a risk that the evaluation work becomes too demanding in resources in relation to the rest of the work in the process. It can therefore simplify things considerably if the person in charge of evaluation has experience of this kind of work and in more complex situations has also preferably research experience. In many organisations, there are internal skills in evaluation work among those for example who work with quality work. Otherwise, such skills must be recruited externally. It is important that evaluation skills are brought into play at an early stage in order to design and integrate this subprocess in the work of the main process.

There are various reasons for carrying out an evaluation, there are different things to evaluate and there are many ways of doing it. We have no space here to describe this special area in more

Chapter 12

detail, but the following questions may help to guide us in reflecting further about this subject:

- Why should we carry out an evaluation?
- For whom are we doing it?
- What should we evaluate?
- How is it to be done?
- Who should do it?

Summary

Health promotion is a form of work dedicated to bringing about change which aims at promoting the preconditions for people's health. The setting or situation for this change can, for example, be a workplace and an organisation. The setting consists of a structure within which various processes take place. There are two kinds of processes, described respectively as *technical ('hard')* and *human-related ('soft')*. It is primarily the human-related processes which have to be influenced in the work of change and which largely determine if there will be a change/movement in the direction of increasingly better preconditions for health in the workplace.

The human-related processes consist of people's ideas, thoughts, personal experiences, attitudes, behaviour and interaction and therefore are incapable of being programmed or steered.

Moldaschl and Brödner[261] describe the organisational change towards greater sustainability as a risky endeavour with a moving target, involving a risk of conflicts and unwanted side effects:

> ..the process of organisational transformation towards sustainable work systems turns out to be a risky endeavour with a moving target, insufficient knowledge, partially conflicting interests and, hence, with wide ranges of uncertainty and unexpected side effects. Thus, our heuristics of reflexive design

and intervention for sustainable change are not a toolbox... they insist on the necessity to introduce knowledge and values, models and strategies. (p. 187).

The health promotion process entails the same form of change that these researchers describe. It is principally a human-related process which is not amenable to any one-type, standard solution. In order to deal with this, *values, strategies, guiding models* and *knowledge* are required.

Knowledge and competence are a crucial basic requirement for intervening in complicated processes of change. We have discussed what theoretical knowledge is needed for health promotion and human-related processes, earlier in the book. In addition, experience-based knowledge of health and work dedicated to change in organisations is needed. These two knowledge bases, theoretical knowledge and experience-based practical know-how, together provide the competence necessary for this kind of work. These skills can partly be obtained externally but it is necessary that the organisation itself also has or tries to establish a competence in health promotion dedicated to development and change. In the long-term, health promotion is only to a small extent a task for consultants: it ought instead to be carried out as an internal process within the company relying upon its own skills, as far as this is possible. (Evaluation on the other hand can be an exception to this rule).

Guiding models, some of which have been presented above, are general examples which help to create order and give the human-related processes a structure. This kind of model facilitates analysis and planning, and is above all a support for management in the process of change.

Chapter 12

Moldaschl and Brödner also discuss values and strategy as important constituents in their model of change. Health promotion, as it has been presented in this book, has its value-basis in the idea of salutogenesis and the four critical components. Together these provide health promotion with a value-based starting point and a direction for the journey and thereby make it a strategy, a possible choice of approach in bringing about a change that will lead to better conditions for health.

Afterword

This book has dealt with how to influence human health and survival in a sustainable society. An important aspect of this is to develop ways of working which function well and without detriment to the health of the people involved. An emphasis above all on humanistic values will also lead to material dividends. Those workplaces which provide job-satisfaction and contribute to people's well-being will also probably be those where people's performance, creativity and belief in the future will be enhanced.

We have put forward a theoretical model which is intended to serve as a framework and a guide.

- The starting point is the idea of salutogenesis, which seeks the origin of health
- Antonovsky's sense of coherence theory (SOC) is used as a guide in discovering *what* is good for health.
- The four criteria -or critical components - form a theory about *how* a change to better conditions for health can take place.

The model presented is only a guide, not a fully-fledged program of action. People are more important than models, and health promotion, in particular, depends on principles which allow people and organisations to increase their own control over those factors influencing health. Health promotion cannot function as something designed to steer and control. In that case, it is not health promotion and it is unlikely that it will lead to any notable improvements in health either. When ideas and theory encounter reality, reality- that is to say the people who make it up- must largely determine what the plan for improved health should include, and how it should be implemented. An initial step is therefore to create trust and a shared understanding by asking questions, and by listening and responding to those ideas which people advocate.

The person who is going to work with health issues in organisations has reason to continue to raise questions and reflect about the following questions: what is health? What is health promotion? What's so special about health promotion? What is the goal - what do we hope to achieve - with the health work which is planned? In the individual workplace, there are also other issues to think about: What is the specific situation here? What local knowledge, views and previous experience can we draw upon to influence the work of health promotion? Is health promotion the right strategy or, rather, how can health promotion be applied to this particular workplace?

It is also important to think about the results which health promotion can bring about and to critically examine if there are unexpected, perhaps undesirable, effects from this form of health work.

Reflecting about goals, methods of working, and also about the outcome of what is done is important if a field is to develop further. Health promotion is a relatively new field with a great deal of development ahead of it.

This continued development will occur in two ways. Partly it will take place through a new theory being put forward. However, above all, it will occur in the practical field through the labours of various professional groups in trying out new approaches in their attempts to make the workplace an environment which promotes health. Through the combined efforts of these groups, and by documenting and studying together with researchers what is done in practice, there can be a positive development of methods for improving health in working life.

The ideas and creativity which drive these processes, benefit from the fact that several perspectives and disciplines are brought together face to face in trying to solve concrete problems. There is a need for more encounters where medically trained health workers and representatives of vocational and social medicine can engage in constructive dialogue with occupational psychologists, economists, educators, people working with fitness and wellness programs and, last but not least, with the other representatives of working life who have the job of incorporating and integrating the ideas and skills associated with health promotion, in their own sphere of activity.

The extensive publication of new books and articles in more and more disciplines points to a ongoing widening of interest. There are numerous research people who wish to take part in investigating and explaining why working life is the way it is and what one can do to improve it,

If the reader of this book has opinions about its content or has constructive criticism to offer, they are welcome to get in touch with the author.

anders@halsopromotiongruppen.se

Bibliography

Abrahamsson, K., Bradley, G. m.fl (2003). *Friskfaktorer i arbetslivet.* Stockholm: Prevent.

Ahrenfelt, B. (1995). *Förändring som tillstånd.* Lund: Studentlitteratur.

Aldana, S.G. Financial Impact of Health Promotion Programs: A Comprehensive Review of the Literature. *American Journal of Health Promotion.* 2001;15(5): 296–320.

Antonovsky, A. (1967). Social class, life expectancy and overall mortality. *Millbank Mem. Fund. Q.* 43:31–73.

Antonovsky, A. (1971). Social and Cultural factors in Coronary Disease: An Israel-North America Sibling Study. *Israel Journal of Medical Sciences,* 7, 1578–1583.

Antonovsky, A. (1979). *Health, Stress and Coping.* San Francisco: Jossey-Bass.

Antonovsky, A. The Sense of Coherence as a Determinant of Health. In J.D. Matarozzo and others (eds.) *Behavioural Health: a Handbook of Health Enhancement and Disease Prevention.* New York: Wiley, 1984.

Antonovsky, A. (1987). *Unraveling the mystery of health: How people manage stress and stay well.* San Francisco: Jossey-Bass.

Antonovsky, A. (1992). Can attitudes contribute to health? *Advances: The Journal of Mind-Body Health,* 8, 33–49.

Antonovsky, A. (1993). The Structure and Properties of the Sense of Coherence Scale. *Social Science Medicine,* 36:6, 725–733.

Antonovsky, A. A sociological Critique of the "Well-Being" Movement. *Advances: The Journal of Mind-Body Health* Vol.10, No. 3, 1994.

Antonovsky, A. (1996).The salutogenic model as a theory to guide health promotion. *Health Promotion International.* Vol. 11 No 1, 11–18.

Arbetslivsinstitutet. 1999:22. *Ett friskt arbetsliv.*

Argyris, C. & Schön, D.A. (1978). *Organizational Learning, a Theory of Action Perspective.* Reading. Mass.: Addison-Wesley.

Arvonen, J. (1989). Att leda via idéer. Lund: Studentlitteratur.

Bateson, G. (1972). Steps to an Ecology of Mind. New York: Ballantine.

Bern, L. (2002). Humankapitalisten-den nya ekonomins professionella aktör. Falun: Ekerlids förlag.

Bjerlöv, M. Deutero-learning and sustainable change. I: Docherty, P., Forslin, J. & Shani, R. (eds.) (2002). Creating Sustainable Work Systems. London: Routledge.

Blaxter, M.(1990). Health and Lifestyles. London: Routledge

Bolman,L. & Deal, T. (2003). Reframing Organizations - Artistry, Choice and Leadership. Jossey-Bass Inc.

Borgenhammar, E. (1993). Att vårda liv. Organisation, etik, kvalitet. SNS Förlag.

Bowman, B. (1996). Cross-cultural validation of Antonovsky's Sense of Coherence Scale. Journal of Clinical Psychology. Vol. 52:2, 547–549.

Brylde, B. & Tengland, P-A. (2003). Hälsa och sjukdom, en begreppslig utredning. Lund: Studentlitteratur.

Bunton, R. & Macdonald, G. (1995). Health Promotion, Disciplines and Diversity. London: Routledge.

Carlsson, G. & Arvidsson, O. (red) (1994). Kampen för folkhälsan. Stockholm: Natur och Kultur.

Collins, J. & Porras, J. (1997). Built to Last, successful habits of visionary companies. New York: Harper business.

Demmer, H. Worksite Health Promotion: How to go about it. WHO: European Health Promotion Series No 4, 1995. ISBN 3-924379-40-8

Docherty, P., Forslin, J. & Shani, R. (eds.) (2002). Creating Sustainable Work Systems. London: Routledge.

Downie, R.S., Tannahill, C. & Tannahill, A. (1996). Health Promotion, Models and Values. Oxford: Oxford University Press.

Edvinsson, L. (1998). Det intellektuella kapitalet. Malmö: Liber ekonomi.

Ekvall, G. (1991). Förnyelse och Friktion, om organisation, kreativitet och innovation. Stockholm: Natur och Kultur.

Ellström, P-E., Gustavsson, B., Larsson, S. (1996). Livslångt lärande. Lund: Studentlitteratur.

Eriksson, K. (1984). Hälsans Idé. Stockholm: Almqvist & Wiksell.

European network for workplace health promotion. The Lisbon Statement on Workplace Health in Small and Medium-Sized Enterprises (SMEs). www.enwhp.org

Feldt, T. (1997). The role of sense of coherence in well-being at work: analysis of main and moderator effects. Work & Stress Vol. 11 No. 2, 134–147.

Folkhälsoinstitutet, Hälsofrämjande skola – ett samlat grepp för visionen om skolan som en stödjande miljö för hälsan, Stockholm: Folkhälsoinstitutet 1997:39.

Forsberg, E. & Starrin, B. (red.) (1997). Frigörande kraft. Förlagshuset Gothia.

Frankl,V. (1984). Man's Search for Meaning. Washington Square Press.

Frenz, A., Carey, M. & Jorgensen, R. (1993). Psychometric Evaluation of Antonovsky's Sense of Coherence Scale. Psychological assessment. Vol. 5 No. 2 145–153.

Geyer, S. (1997). Some conceptual considerations on the Sense of Coherence. Social Science Medicine .Vol. 44:12, 1771–1779.

Health Education Authority (1998). Working with small and medium-sized enterprises. Summary report. HEA London.

Herzberg, F., Mausner, B. & Snyderman, B. (1959). The motivation to Work. New York: Wiley & Sons.

Huzzard,T. The Convergence of the Quality of Working Life and Competitiveness. A Current Swedish Literature Review. Stockholm: Arbetslivsinstitutet 2003:9.

Ingvar, D H. & Sandberg, C.G. (1991). Det medvetna företaget. Stockholm: Timbro.

Isaac, F. & Flynn, P. Johnson & Johnson LIVE FOR LIFE Program Now and Then. American Journal of Health Promotion 2001;15(5):365–367.

IVA (2000). Ett friskt arbetsliv – Humankapitalets strategiska betydelse för företag. Stockholm: IVA ISSN 1102-8254.

Johansson, O. (1990). Organisationsbegrepp och begreppsmedvetenhet. Göteborg: Bas ekonomisk förening.

Johnsson, J., Lugn, A. & Rexed, B. (2003). Långtidsfrisk – Så skapas hälsa, effektivitet och lönsamhet. Ekerlids Förlag.

Karasek, R. & Theorell, T. (1990). Healthy Work. New York: Basic Books Inc.

Kickbusch, I. (1997). Think health: what makes the difference? Health Promotion International. Vol 12, No4 (265–272).

Kira, M. (2003). From Good Work to Sustainable Development- Human Resources Consumption and Regeneration in the Post-Bureaucratic Working Life. A dissertation submitted to the Royal Institute of Technology Stockholm: KTH.

Kirsten, Wulf. Health and Productivity Management – a Future Model for Europe? Newsletter nr 4:2003. www.enwhp.org

Klockars, K. & Österman, B. (red) (1995). Begrepp om hälsa. Filosofiska och etiska perspektiv på hälsa och livskvalitet. Stockholm: Liber.

Kobasa, S.C., Maddi, S.R. & Kahn, S. (1982). Hardiness and Health: A prospective study. Journal of Personality and Social Psychology, 42, 168–177.

Korp, P. (2002). Hälsopromotion- en sociologisk studie av hälsofrämjandets institutionalisering. Dissertation. Sociological Institute. University of Gothenburg.

Koskinen, L. (1989). Etikens teori i praktiken. i Etikens pris. Antologi. Stockholm: Verbum.

Kuhn, T.S. (1970) The Structure of Scientific Revolutions. The University of Chicago Press.

Kylén, S. (1993). Arbetsgrupper med förändrings och utvecklingsuppdrag, från defensiva till offensiva rutiner. Göteborgs universitet, Psykologiska inst.

LaLonde, M (1974) A new Perspective on the Health of Canadians - a working document. Cat No. H31-1374 ISBN 0-662-50019-9

Leithwood, K., Jantzi, D. & Steinbach, R. (2002). Changing Leadership for Changing Times. Philadelphia: Open University Press.

Lennéer-Axelsson, B. & Thylefors, I. (1996). Om konflikter, hemma och på jobbet. Stockholm: Natur och Kultur.

Lindberg, P. & Vingård, E. De friska. In the Swedish government official report SOU 2002: Handlingsplan för ökad hälsa i arbetslivet .

Linell, P. (1978). Människans språk. Lund: Liber Läromedel.

Liss, P-E. & Petersson, B. (red) (1995). Hälsosamma tankar. 11 filosofiska uppsatser tillägnade Lennart Nordenfeldt. Nora: Nya Doxa

Maltén, A. (1981). Vad är kunskap? Malmö: Gleerups förlag.

Maltén, A. (2000). Det pedagogiska ledarskapet. Lund: Studentlitteratur.

Marmgren, L. Att leda mjuka processer. MiLs Skriftserie nr 17/2002. ISSN 1402-3466.

Martin Lindberg, Tillväxt förutsätter uthålligt arbetsliv, LO-tidningen 2004:1.

Nutbeam, D. & Harris, E. (1999). Theory in a Nutshell, a Guide to Health Promotion. Australia, McGraw-Hill.

Nutbeam, D. Health Promotion Glossary. Health Promotion Int. 1998:13, 14 (349–364).

Ohlson, C-G., Sandberg, C-G. (1987). The bio-psycho-social model in industrial medicine and management. Stockholm: Arbetsmiljöfonden.

Olsson, U. (1997). Folkhälsa som pedagogiskt projekt. Avhandling, Uppsala Universitet.

Palmblad, E. (1989). Medicinen som samhällslära. Göteborg: Daidalos.

Palmblad, E., Eriksson, B.E. (1995). Kropp och politik. Stockholm: Carlssons.

Peterson, B. et al. (1995). Hälsosamma tankar. Nora: Nya Doxa.

Poland, B., Green, W. & Rootman (red) (2000). Settings for Health Promotion, Linking Theory and Practice. London: Sage Publications.

Prochaska, J.O. & DiClemente, C.C. Stages and processes of self-change of smoking: Toward an intergrative model of change. Journal of Consult. Clin. Psychology. 1983; 51:390–395.

Qvarsell, R. & Torell, U. (red) (2001). Humanistisk hälsoforskning – en forskningsöversikt. Lund: Studentlitteratur.

Rendahl, J-E., Hart, H., Lawler, E., Ledford, G. & Norrgren, F. (1995). Att förändra och leda morgondagens arbete. Stockholm: Vis strategi AB.

Rimann, M. & Udris, I. Kohärenzerleben: Zentraler Bestandsteil von Gesundheit oder Gesundheitsressource? Sid. 351 i: Schuffel et. al (red.) (1998). Handbuch der Salutogenese. Wiesbaden: Ullstein Medical Verlagsgesellschaft.

Rydqvist, L. & Winroth, J. (2003) Idrott, friskvård, hälsa och hälsopromotion. Farsta: SISU idrottsböcker.

Sachs, L., Uddenberg, N. (1988). Medicin, myter, magi. Stockholm: Natur och Kultur.

Sandberg, J. & Targama, A. (1998). Ledning och förståelse, ett kompetensperspektiv på organisationer. Lund: Studentlitteratur

Sandkull, B. & Johansson, J. (1996). Från Taylor till Toyota. Lund: Studentlitteratur.

Schein, E. H. (1988). Process Consultation Volume I – Its Role in Organization Development. New York: Addison-Wesley Publishing Company.

Seedhouse, D. (2004). Health Promotion, Philosophy, Prejudice and Practice. Chicester: Wiley & Sons.

Seligman, M.E. & Garber, J. (eds) (1980). Human helplessness, theory and applications. New York: Academic Press.

Selye, H. (1956). Stress of Life. New York: McGraw-Hill.

Sennett, R. (1998). The Corrosion of Character, the personal consequences of work in the new capitalism. New York: W. W. Norton & Company Inc.

Shook, R. (1990). Turnaround: The new Ford Motor Company. New York: Prentice-Hall.

Sivik, T. & Theorell, T. (red.) (1995). Psykosomatisk medicin. Lund: Studentlitteratur.

SOU 2002: 5 Handlingsplan för ökad hälsa i arbetslivet.

Starrin, B. & Jönsson, L.R. Ekonomisk påfrestning, skamgörande erfarenheter och ohälsa under arbetslöshet. Arbetsmarknad & Arbetsliv, årg 4:2 1998.

Steinberg, J.M. (1994). Den nya inlärningen. Stockholm: Ekerlunds.

Svedberg, L. (2003). Gruppsykologi, om grupper, organisationer och ledarskap. Lund: Studentlitteratur.

Tamm. M. (1996). Modeller för hälsa och sjukdom. Liber: Malmö.

Taylor, F.W. (1967). Principles of Scientific Management. New York: WW Norton&Co Ltd

The Luxembourg Declaration on Workplace Health Promotion in the European Union.(1997) www.enwhp.org

Their, S. (1996). Det pedagogiska ledarskapet. Mariehamn: Mermerus förlag.

Thelander, E. (2003). Delaktighet och dialog – på väg mot hållbara arbetsplatser. Stockholm: Arbetslivsinstitutet.

Theorell, T. Är ökat inflytande på arbetsplatsen bra för folkhälsan? Kunskapssammanställning. Statens folkhälsoinstitut. 2003:46.

Tishelman, C. (1996). Critical Reflections over the uncritical use of Antonovsky's Sense of Coherence Questionnaire. Vård i Norden 1, 33–37.

Tones, B.K. & Tilford, S. (1994). Health Education: Effectiveness, Efficiency and Equity. London: Chapman & Hall.

Trollestad, C. (1998). Människosyn i ledarskapsutveckling. Nora: Nya Doxa.

Watzlawick, P. Weakland, J & Fish, R. (1974). Change; Principles of Problem Formation and Problem Resolution. W.W. Norton&Company

Wenzel, E. A comment on settings in health promotion. Internet Journal of Health Promotion, 1997.

Whitelaw, S., Baxendale, A., Bryce, C., Machardy, L., Young, I. & Witney, E. (2001). Settings based health promotion: a review. Health Promotion International, Vol.16: 4, 339–353

WHO, Proceedings and final acts of the international health conference 19 June-22 July 1946 WHO interim commission. www.who.int/library/collections/historical/

WHO Constitution (1948). www.who.int/library/collections/historical/

WHO Statement on Healthy Workplaces (1997) Fourth International Conference on Health Promotion.

World Health Organization. (1978). Declaration of Alma-Ata. http://www.euro.who.int/AboutWHO/Policy/20010827_1

WHO, Ottawa Charter for Health Promotion (1986) www.who.int

WHO, Supportive Environment for Health (1991). Sundsvall. www.who.int

WHO, Fourth International Conference on Health Promotion, Jakarta 21–25 July 1997. www.who.int

WHO: report, 2002: Good Practice in Occupational Health Services – A Contribution to Workplace Health.

WHO, HEALTH21: an introduction to the health for all policy framework for the WHO European Region.(latest update 2005) www.euro.who.int

von Otter (red). Ute och inne i svenskt arbetsliv – forskare analyserar och spekulerar om trender i framtidens arbetsliv. Arbetslivsinstitutet, Arbetsliv i omvandling 2003: 8.

Wynne, R. & Clarkin, N. (eds.) (1992). Under construction. Building for Health in the EC Workplace. Luxembourg: Office for Official Publications of the European Communities.

Wynne, R. (1993). Action for Health at Work, the Next Steps. Dublin: European Foundation for the Improvement of Living and Working Conditions. (Working paper EF/93/19/EN/DE/FR)

Wynne, R. (1997). A manual for training in workplace health promotion. European Foundation for the Improvement of Living and Working Conditions.

Footnotes

[1] WHO, HEALTH21: an introduction to the health for all policy framework for the WHO European Region. (latest update 2005). www.euro.who.int

[2] Lindberg, P. & Vingård, E. De friska. In the Swedish government official report SOU 2002: Handlingsplan för ökad hälsa i arbetslivet.

[3] SOU 2002:5 Handlingsplan för ökad hälsa i arbetslivet.

[4] IVA (2000). *Ett friskt arbetsliv- Humankapitalets strategiska betydelse för företag.* Stockholm: IVA

[5] Collins,J. & Porras, J(1997). *Built to Last, successful habits of visionary companies,* New York: Harper business.

[6] Shook, R. (1990). *Turnaround: The new Ford Motor Company.* New York: Prentice-Hall.

[7] IVA (2000). *Ett friskt arbetsliv. Humankapitalets strategiska betydelse för företag.*

[8] Edvinsson, L. (1998). *Det intellektuella kapitalet.*

[9] Bern, L. (2002). *Humankapitalisten- den nya ekonomins professionella aktör.* Falun: Ekerlids förlag.

[10] Sennett, R. (1998). *The Corrosion of Character, the personal consequences of work in the new capitalism.* New York: W.W.Norton & Company Inc.

[11] SOU 2002:5 Handlingsplan för ökad hälsa i arbetsliv.

[12] See: Arbetslivsinstitutet 1999:22, *Ett friskt arbetsliv.*

[13] Seedhouse, D. (2004). *Health Promotion, Philosophy, Prejudice and Practice.* Wiley & Sons.

[14] Borgenhammer, E. (1993). *Att vårda liv. Orgamisation, etik, kvalitet.* SNS. Förlag.

[15] Antonovsky, A. (1992). *Hälsans mysterium.* Stockholm: Natur och Kultur

[16] Antonovsky, A. (1979). *Health, Stress and Coping,* San Francisco: Jossey-Bass.

[17] Sachs,L. & Uddenberg,N. (1988). *Medicin, myter, magi.* Natur och Kultur.

[18] Borgenhammer, E. (1993). *Att vårda liv. Organisation, etik, kvalitet.* SNS Förlag.

[19] Sivik, T. & Teorell,T. (1995). *Psykosomatisk medicin.* Lund: Studentlitteratur.

[20] Selye, H.(1956). *Stress of Life.* New York: McGraw-Hill.

[21] Lalonde, M.(1974). *A New Perspective on the Health of Canadians.* ISBN 0-662-50019-9

[22] Peterson, B. et al. (1995). *Hälsosamma tankar.* Nora: Nya Doxa.

23 Carlssson & Arvidsson(ed.) (1994). *Kampen för folkhälsan*. Stockholm, Natur och Kultur.
24 Bonniers lexikon (1996).
25 Olsson, U. (1997). *Folkhälsa som pedagogiskt projekt*. Dissertation Uppsala University.
26 Carlsson & Arvidsson (eds) (1994). *Kampen för folkhälsan*. Stockholm, Natur och Kultur.
27 Palmblad, E. & Eriksson, B.E. (1995). *Kropp och politik*. Stockholm: Carlssons.
28 Olsson, U. (1997). *Folkhälsa som pedagogiskt projekt*. Avhandling Uppsala Universitet.
29 Lalonde, M.(1974). *A New Perspective on the Health of Canadians*. ISBN 0-662-50019-9
30 World Health Organization. (1978). *Declaration of Alma-Ata*. www.who.dk/aboutWHO/Policy.
31 Antonovsky, A. (1979). Health, Stress and Coping. Jossey-Bass, San Francisco.
32 SOU 2002: *Handlingsplan för ökad hälsa i arbetsliv*.
33 The concept of *long-term healthy* was proposed in 1990 at the timber company Stora Enso in Fors by the company doctor Johnny Johnsson. A person is longterm healthy if they have not had a single day when they have been absent due to illness during a two year period.
34 Borgenhammar, E. (1993). *Att vårda liv. Organisation, etik, kvalitet*. SNS Förlag. Borgenhammar speak of *tillitsbristsjukdomar* i.e. *illnesses due to lack of trust*. However, *alienation syndrome* perhaps better suggests what is really involved.
35 Downie, R.S. Fyfe, C. & Tannahill, A. (1991). *Health Promotion. Models and Values*. Oxford University Press.
36 Klockars,K. & Österman, B. (red) (1995). *Begrepp om hälsa. Filosofiska och etiska perspektiv på hälsa och livskvalitet*. Stockholm: Liber.
37 Gadamer in Brylde,B. & Tengland,P.-A. (2003). *Hälsa och sjukdom, en begreppslig utredning*. Lund: Studentlitteratur.
38 Antonovsky, A. (1979). *Health, Stress and Coping*. San Francisco, Jossey-Bass.
39 Sivik,T. & Teorell,T. (1995). *Psykomatisk medicin*. Lund, Studentlitteratur.
40 Eriksson, K. (1984). *Hälsans Idé*. Stockholm: Almqvist & Wiksell.
41 Rydqvist,L.&Winroth, J. (2002). *Idrott, friskvård, hälsa och hälsopromotion*. SISU: idrottsböcker.
42 Selye, H. (1956). *Stress of Life*. New York, MccGraw-Hill.
43 Antonovsky, A. (1987) *Unraveling the Mystery of Health*. San Francisco-London: Jossey-Bass Publishers.

44 Since 1996, the program in Health Promotion at the University West www.hv.se.
45 WHO Constitution (1948). It is available at the web page www.who.int.
46 Downie, R.S., Tannahill, C. & Tannahill, A. (1996). *Health Promotion, Models and Values.* Oxford University Press.
47 LaLonde, M (1974) A new Perspective on the Health of Canadians - a working document. Cat No. H31-1374 ISBN 0-662-50019-9
48 Lalonde, *Health Promotion Strategy*, ww.hc-sc-gc.ca.
49 WHO, Ottawa *Charter for Health Promotion* (1986) www.who.int
50 WHO, *Ottawa Charter for Health Promotion* (1986) www.who.int
51 Blaxter, M.(1990). *Health and Lifestyles.* London: Routledge.
52 Antonovsky, A. (1987) *Unravelling the Mystery of Health.* San Francisco-London: Jossey-Bass Publishers.
53 WHO, Ottawa *Charter for Health Promotion* (1986) www.who.int.
54 WHO, Fourth International Conference on Health Promotion, Jakarta 21-25 July 1997. www.who.int.
55 The Luxembourg Declaration on Workplace Health Promotion in the European Union.(1997) www.enwhp.org
56 European Foundation http:/www.eurofound.ie/
57 Wynne, R. & Clarkin, N. (eds) (1992). *Under construction. Building for Health in the EC Workplace.* Luxembourg: Office for Official Publications of the European Communities.
58 Wynne, R. (1993). *Action for Health at Work, the Next Steps.* Dublin: European Foundation for the Improvement of Living and Working Conditions. (Working paper EF/93/19/EN/DE/FR)
59 www.enwhp.org
60 Bundesverband der Betriebskrankenkassen www.bkk.de.
61 www.enwhp.org
62 Occupational Health and Safety.
63 www.enwhp.org
64 SME= small and medium sized entreprises.
65 www.enwhp.org
66 www.enwhp.org
67 See e.g. Qvarsell,R. & Torell, U. (eds) (2001). *Humanistisk hälsoforskning- en forskningsöversikt.* Lund : Studentlitteratur.
68 See page 179, in Docherty, P. Forslin, J. & Shani, R. (eds.) (2002). *Creating Sustainable Work Systems.* London: Routledge.
69 Watzlawick, P. Weakland, J & Fish, R. (1974). *Change; Principles of Problem Formation and Problem Resolution.* W.W. Norton&Company

70 Collins, J.C. & Porras, J. I (1997). *Built to last, Successful Habits of Visionary Companies.* New York: Harper business.

71 Bolman,L. & Deal, T. (2003). *Reframing Organizations - Artistry, Choice and Leadership.* Jossey-Bass Inc.

72 Antonovsky, A. (1993). The Structure and Properties of the Sense of Coherence Scale. *Social Science Medicine,* 36:6, 725-733.

73 Ellström, P-E. Gustavsson, B. Larsson, S. (1996). *Livslångt lärande.* Lund: Studentlitteratur.

74 Antonovsky, A. A sociological critique of the "Well-Being" Movement. *Advances: The Journal of Mind-Body Health* Vol. 10, No. 3, 1994.

75 Ellström, P-E. Gustavsson, B. Larsson, S (1996). *Livslångt lärande.* Lund: Studentliyyeratur.

76 In Rendahl,J.E. et al. (1995). *Att förändra och leda morgondagens arbete.* Stockholm: Vis strategi AB.

77 Sennet, R. (1998). *The Corrosion of Character, the personal consequences of work in the new capitalism.* New York: W. W. Norton & Company Inc.

78 Antonovsky, A. (1996). The salutogenic model as a theory to guide health promotion. *Health Promotion International.* Vol 11 No. 1,11-18.

79 Kuhn, T.S. (1970) *The Structure of Scientific Revolutions.* The University of Chicago Press.

80 Antonovsky, A. (1979). *Health, Stress and Coping.* San Francisco :Jossey-Bass.

81 Feldt,T. (1979). The role of the sense of coherence in well-being at work: analysis of main and moderate effects. *Work & Stress* Vol. 11 No. 2, 134-147.

82 Franz, A. Carey, M. & Jorgensen, R. (1993). Psychometric Evaluation of Antonovsky's Sense of Coherence Scale. *Psychological assessment* Vol. 5 No. 2. 145-153.

83 Tishelman, C. (1996). Critical Reflections over the uncritical use of Antonovsky's Sense of Coherence Questionnaire. *Vård i Norden.*

84 Antonovsky, A. (1967). Social class, life expectancy and overall mortality. Millbank Mem. Fund. Q. 43:31-73.

85 Antonovsky, A. (1979). *Health, Stress and Coping.* Jossey-Bass, San Francisco.

86 Antonovsky, A. (1996). The salutogenic model as a theory to guide Health Promotion. *Health Promotion International,* 11, 11-18.

87 Antonovsky, A. (1971). Social and Cultural factors in Coronary Disease: An Israel-North America Sibling Study. *Israel Journal of Medical Sciences,* 7, 1578- 1583.

88 Antonovsky, A. (1979). *Health, Stress and Coping.* San Francisco: Josey-Bass. (p. 184).

[89] Antonovsky, A. (1996). "The Salutogenic Model as a theory to guide Health Promotion. Health Promotion International, 11, 11-18.

[90] Antonovsky, A. "The Sense of Coherence as a Determinant of Health." In J.D. Matarozzo and others (eds.) *Behavioral Health: a Handbook of Health Enhancement and Disease Prevention.* New York: Wiley, 1984.

[91] Antonovsky, A. (1987) *Unravelling the Mystery of Health, How People Manage Stress and Stay Well.* San Francisco-London. Jossey-Bass.

[92] Karasek, R. & Theorell, T. (1990). *Healthy Work.* New York : Basic Books Inc.

[93] Kobasa, S.C. Maddi, S.R. & Kahn, S. (1982). Hardiness and Health: A prospective study. *Journal of Personality and Social Psychology*, 42, 168-177.

[94] Frankl,V. (1984). *Man´s Search for Meaning.* Washington Square Press.

[95] (with Nietzsche´s words: "Wer ein Warum zu leben hat, erträgt fast jedes Wie.")

[96] Antonovsky, A. (1992). Can attitudes contribute to health? *Advances, The Journal of Mind- Body Health*, 8, 33-49.

[97] Antonovsky, A. (1987), *Unraveling the mystery of health: How people manage stress and stay well.* San Francisco : Jossey-Bass.

[98] See inter alia:
Antonovsky, A. (1993). The Structure and Properties of the Sense of Coherence Scale. *Social Science Medicine*, 36:6, 725-733.
Bowman, B (1996). Cross-cultural validation of Antonovsky's Sense of Coherence Scale. *Journal of Clinical Psychology* Vol. 52:2, 547-549.
Feldt, T. (1997). The role of sense of coherence in well-being at work: analysis of main and moderator effects. *Work & Stress*, 11.2, 134-147.

[99] Geyer,S. (1997). Some conceptual considerations on the Sense of Coherence. *Social Science Medicine* Vol. 44:12, 1771-1779.

[100] Tishelman, C.(1996). *Critical reflections over the uncritical use of Antonovsky's Sense of Coherence questionnaire.* Vård i Norden 1, 33-37.

[101] Nutbeam,D. Health promotion Glossary. *Health Promotion Int.* 1998:13,14 (349-364).

[102] Collins, J.C. & Porras, J.I. (1994). *Built to last. Successful habits of Visionary Companies.* New York: Harper Business.

[103] Sennet, R. (1998). *The Corrosion of Character, the personal consequences of work in the new capitalism.* New York; W.W.Norton & Company Inc.

[104] Ingvar, D.H. &Sandberg, C.G. (1991). *Det medvetna företaget.* Stockholm: Timbro.

[105] Folkhälsoinstitutet, *Hälsofrämjande skola- ett samlat grepp för visionen om skolan som en stödjande miljö för hälsan.* Stockholm: Folkälsoinstitutet 1997:39.

[106] Sandberg, J. & Targama, A. (A) (1998). *Ledning och förståelse, ett kompetensperspektiv på organisationer.* Lund: Studentlitteratur.

[107] Ohlson, C-G, Sandberg CG. *The bio-psycho-social model in industrial medicine and management.* Stockholm: Arbetsmiljöfonden, 1987.

[108] The word traditional underlines the fact that the rehabilitative work which starts with an injury or functional handicap represents a traditional viewpoint, not a salutogenic one. Moreover a rehabilitation group can be an object for salutogenic health promotion where the point of departure is the functional capacity and the resources which, despite everything, are to be found in this category of people.

[109] See e.g. Palmblad, E. (1989). *Medicinen som samhällslära.* Göteborg: Daidalos.

[110] European Network for Workplace Health Promotion www.enwhp.org

[111] The Luxembourg Declaration on Workplace Health Promotion in the European Union.(1997) www.enwhp.org

[112] Karasek, R. & Theorell, T. (1990). *Healthy Work.* New York : Basic Books Inc.

[113] Ekvall, G. (1991). *Förnyelse och Friktion, om organisation kreativitet och innovation.* Stockholm: Natur och Kultur.

[114] Isaac, F. & Flynn, P. Johnson & Johnson LIVE FOR LIFE Program Now and Then. *American Journal of Health Promotion* 2001; 15(5):365-367.

[115] European Network for Workplace Health promotion. www.enwhp.org.

[116] Wynne, R. (1997). *A manual for training in workplace health promotion.* European Foundation for the Imrovement of Living and Working Conditions.

[117] Wynne,R. & Clarke,N. (1993). *Under construction: Building for health in the EC workplace* Office for official publications of the European Communities, Luxemburg.

[118] http://www.eurofound.eu.int/pubdocs/1997/24/en/1/ef9724en.pdf

[119] Bunton, R. & Macdonald, G. (1995). *Health Promotion, Disciplines and Diversity,* London: Routledge.

[120] Epidemiology is the study of the incidence and distribution of diseases, and of their control and prevention.

[121] Koskinen, L. (1998). *Etikens teori i praktiken.* I *Etikens pris,* Anthology. Stockholm: Verbum.

[122] Trollestad, C. (1998). *Människosyn i Ledarskapsutveckling:* Nya Doxa.

[123] Tones. B.K. & Tilford, S. (1994). *Health Education: Effectiveness, Efficiency and Equity.* London: Chapman & Hall.

[124] Prochaska, J.O. & DiClemente. C.C. Stages and processes of self-change of smoking: Towards an integrative model of change. *Journal of Consult. Clin. Psychology.* 1983; 51::390-395.

[125] Nutbeam, D. & Harris, E. (199). *Theory in a Nutshell a Guide to Health promotion.* McGraw-Hill Australia.

[126] In Bunton, R. & Macdonald, G. (1995). *Health Promotion, Disciplines and Diversity.* London: Routledge.

[127] See e.g. Korp, P. (2002). Hälsopromotion- en sociologisk studie av hälsofrämjandets institutionalisering. Dissertation. Sociological Institute. University of Gothenburg.

[128] Bateson, G. (1972). *Steps to an Ecology of Mind.* New York. Ballantine.

[129] Ahrenfelt, B. (1995). *Förändring som tillstånd..* Lund. Studentlitteratur.

[130] Watzlawick, P. Weakland, J & Fish, R. (1974). *Change; Principles of Problem Formation and Problem Resolution.* W.W. Norton&Company

[131] Seedhouse, D. (2004) *Health Promotion, Philosophy, Prejudice and Practice.* Wiley & Sons.

[132] Seedhouse, D. (2004). *Health Promotion, Philosophy, Prejudice and Practice.* Wiley & Sons.

[133] Wynne, R. & Clarkin, N. (eds.). *Under construction. Building for Health in the EC Workplace.* Luxembourg : Office for Official Publications of the European Communities.

[134] See the Swedish official report SOU 2002:5. *Handlingsplan för ökad hälsa i arbetslivet.*

[135] Demmer,H. Worksite Health Promotion: How to go about it. WHO: *European Health Promotion Series* Nr. 4, 1995. ISBN 3-924379-40-8.

[136] Kirstenn, W. Health and Productivity Management- A Future Model for Europe? www.enwhp.org Newsletter nr 4:2003.

[137] See e.g.: Aldana, S.G. Financial Impact of Health promotion Programs: A Comprehensive Review of the Literature . American Journal of Health promotion. 2001; 15(5): 296-320.

[138] Collins, J.C. & Porras, J.L. (1994). *Built to last. Successful Habits of Visionary Comapanies.* New York.: Harper Business.

[139] Docherty, P. Forslin, J. & Shani, R. (eds.) (2002). *Creating Sustainable Work Systems.* London: Routledge.

[140] Johnson,J. Lugn,A. & Rexed, B. (2003). *Långstidsfrisk- så skapas hälsa, effektivitet och lönsamhet.* Ekerlids Förlag.

[141] Leithwood, K. Jantzi, D. & Steinbach, R. (2002). *Changing Leadership for Changing Times.* Philadelphia: Open University Press.

[142] Seedhouse, D. (2004). *Health Promotion, Philosophy, Prejudice and Practice.* Wiley & Sons.

[143] Antonovsky, A. (1996). The Salutogenic Model as a theory to guide Health Promotion. *Health Promotion International,* 11, 11-18.

[144] WHO, Ottawa *Charter for Health Promotion* (1986) www.who.int.

[145] The Luxembourg Declaration on Workplace Health Promotion in the European Union. (1997) (a new version in june 2005) www.enwhp.org

[146] Health Education Authority (1998). *Working with small and medium-sized enterprises.* Summary report. HEA London.

[147] In Docherty, P. Forslin,J. & Shani,R. (eds) (2002). *Creating Sustainable Work Systems.* London: Routledge.

[148] WHO, Ottawa Charter for Health Promotion (1986). See http://www.who.int/entity/en/

[149] The Luxembourg Declaration on Workplace Health Promotion in the European Union. (1997) (a new version in June 2005) www.enwhp.org

[150] Shani, A.B. & Sena,B.J. in Docherty, P. Forslin, J. & Shani, R. (eds) (2002). *Creating Sustainable Work Systems.* London: Routledge.

[151] Liss, P-E. & Petersson, B. (red) (1995). *Hälsosamma tankar 11 filosofiska uppsatser tillägnade Lennart Nordenfeldt.* Nya Doxa.

[152] WHO:report, 2002: Good Practice in Occupational Health. A Contribution to Workplace Health.

[153] Antonovsky, A. (1979). *Health, Stress and Coping.* San Francisco. Jossey-Bass.

[154] Collins, J.C. & Porras, J.I. (1994). *Built to last. Successful Habits of Visionary Companies.* New York: Harper Business.

[155] Brylde,B & Tengland, P-A. (2003). *Hälsa och sjukdom, en begreppslig utredning.* Lund. Studentlitteratur.

[156] Linell, P. (1978). *Människans språk.* Lund: Liber Läromedel.

[157] The Luxembourg Declaration on Workplace Health Promotion in the European Union. (1997) (a new version in June 2005) www.enwhp.org

[158] WHO: report, 2002: Good practice in Occupational Health.

[159] Kickbusch, I. (1997). *Think health: what makes the difference?* Health Promotion International Vol 12, No4 (265-272).

[160] Appears in Poland,B. Green, W. & Rootman (eds) (2000). *Settings for Health promotion, Linking Theorey and Practice.* London: Sage Publications.

[161] Belin, L. in Sivik,T. & Theorell.T. (eds) (1995) *Psykosomatisk medicin.* Lund: Sudentlitteratur.

[162] The Luxembourg Declaration on Workplace Health Promotion in the European Union. (1997) (a new version in June 2005) www.enwhp.org

[163] Kira,M. (2003). *From Good Work to Sustainable Development- Human Resources Consumption and Regeneration in the Post-Bureaucratic Working Life.* A dissertation submitted to the Royal Institute of Technology Stockholm: KTH.

[164] Korp, P. (2002). *Hälsopromotion- en sociologisk studie av hälsofrämjandets institutionalisering.* Dissertation. Sociological Institute. University of Gothenburg.

165. Seedhouse, D. (2004). *Health Promotion, Philosophy, Prejudice and Practice.* Wiley & Sons.
166. Antonovsky, A. (1979). *Health, Stress and Coping.* Jossey-Bass, San Francisco.
167. Antonovsky,A. (1996). The Salutogenic Model as a theory to guide Health promotion. *Health promotion International,* 11, 11-18.
168. Sandkull, B & Johansson,J. (1996). *Från Taylor till Toyota.* Lund: Studentlitteratur.
169. Seedhouse, D. (2004). *Health Promotion, Philosophy, Prejudice and Practice.* Wiley & Sons.
170. Rimann, M. & Udris, I. *Kohärenzerleben: Zentrale Bestandsteil von Gesundheit oder Gesundheitsressource?* p.351 in Schuffel et al. (eds) (1998) *Handbuch der Salutogenese* Wiesbaden: Ullstein Medical Verlagensgesellschaft.
171. WHO, Ottawa Charter for Health Promotion (1986) www.who.int
172. WHO(1998) Health promotion Glossary. Geneva.www.who.int.
173. WHO, Supportive Environment for Health (1991) Sundswall. www.who.int.
174. Wenzel, E. A comment on settings in health promotion. *Internet Journal of Health Promotion,* 1997. www.monash.edu.au/health.
175. Refers here to health education as preventive health work, which is pathogenically based and evaluated. See e.g. Tones,K & Tillford, S. (1994). *Health Education . Effectiveness, Efficiency and Equity.* London: Chapman & Hall.
176. Poland,B. Green,W. & Rootman, I. (2000). *Settings for Health promotion, Linking Theory and Practice.* London: Sage Publications.
177. WHO *Statement on Healthy Workplaces* (1997) Fourth International Conference on Health Promotion.
178. European network for workplace health promotion. The Lisbon Statement on Workplace Health in Small and Medium-Sized Enterprises (SMEs) www.enwhp.org.
179. Bolman,L. & Deal, T. (2003). *Refraiming Organizations - Artistry, Choice and Leadership.* Jossey-Bass Inc.
180. See p80 in Docherty,P. Forslin,J. & Shani, R. (eds.) (2002). *Creating Sustainable Work Systems.* Emerging Perspectives and Practice. London: Routledge.
181. An organization existing without a geographical location, producing only via electronic communication.
182. See for example: Watzlawick, P. Weakland, J & Fish, R. (1974). *Change; Principles of Problem Formation and Problem Resolution.* W.W. Norton&Company
183. Johansson, O. (1990). *Organisationsbegrepp och begreppsmedvetenhet.* Göteborg: Bas ekonomisk förening.

[184] Also called Taylorism efter Frederick Taylor who introduced a systematic division of production into specific stages which were designed so that the workforce, time and other production resources could be employed more efficiently.

[185] Ahrenfelt, B. (1995). *Förändring som tillstånd*. Lund: Studentlitteratur.

[186] Kickbusch in: Whitelaw, S. Baxendale,A. Bryce, C. Machardy, L. Young, I. & Witney,E. (2001). Settings based health promotion: a review. *Health Promotion International*, Vol. 16:4, 339-353.

[187] The Luxembourg Declaration on Workplace Health Promotion in the European Union. (1997) (a new version in June 2005) www.enwhp.org

[188] Included in Docherty,P. Forslin, J. & Shani, R. (eds.) (2002). *Creating Sustainable Work Systems*. London: Routledge.

[189] Collins, J.C. & Porras, J.I. (1994). *Built to last. Successful Habits of Visionary Companies*. New York: Harper Busioness.

[190] The task or production which forms the normal day to day activity of the workplace.

[191] See e.g. Taylor, F.W. (1967). *Principles of Scientific Management*. New York: WW Norton&Co Ltd

[192] Whitelaw, S. Baxendale,A. Bryce, C. Machardy, L. Young, I. & Witney,E. (2001). Settings based health promotion: a review. *Health Promotion International*, Vol. 16:4, 339-353.

[193] IVA(2000). Ett friskt arbetsliv- Humankapitalets strategiska betydelse för företag. Stockholm: IVA. ISSN 1102-8254.

[194] Ellström, P-E, Gustavsson, B. Larsson, S. (1996). *Livslångt lärande*. Lund: Studentlitteratur.

[195] Maltén,A. (2000). Det pedagogiska ledarskapet. Lund: Studentlitteratur.

[196] Kirstenn, W. Health and Productivity Management- A Future Model for Europe? www.enwhp.org Newsletter nr 4:2003.

[197] Whitelaw, S. Baxendale, A. Bryce, C. MacHardy, L. Young, I. &Witney,E. (2001) Settings based health promotion: a review. *Health Promotion International*, Vol. 16:4, 339-353.

[198] Whitelaw, S. Baxendale,A. Bryce, C. MacHardy,L. Young,I. &Witney,E. (2001) Settings based health promotion: a review. *Health Promotion International*, Vol. 16:4, 339-353.

[199] Kylén,S. (1993). *Arbetsrutiner med förändrings och utvecklingsuppdrag, Från defensiva till offensiva rutiner*. Göteborgs universitet, Psykologiska inst.

[200] Bjerlöv, M. Deutero-learning and sustainable change in Docherty, P. Forslin, J. & Shani, R. (eds) (2002). *Creating Sustainable Work Systems*. London: Routledge.

[201] Proceedings and final acts of the international health conference 19 June-22 July 1946 WHO interim commission. http://whqlibdoc.who.int.

[202] WHO's report 2002: *Good Practice in Occupational Health Services- A Contribution to Workplace Health.* www.euro.who.int.
[203] Bateson, G. (1972). *Steps to an Ecology of Mind.* New York: Ballantine.
[204] Argyris,C. & Schön,D.A. (1978). *Organizational Learning. A Theory of Action Perspective.* Reading,Mass. Addison-Wesley.
[205] Karasek, R. & Theorell, T. (1990). *Healthy Work.* New York: Basic Books Inc.
[206] See e.g. the interview with Jan Forslin in the newsletter 2004:1 utveckla.nu.
[207] Docherty,P. Forslin, J. & Shani, R. (eds). (2002). *Creating Sustainable Work Systems.* London: Routledge.
[208] Martin Lindberg, Tillväxt förutsätter uthålligt arbetsliv, LO-tidningen 2004:1.
[209] IVA-M 329. *Ett friskt arbetsliv- Humankapitalets strategiska betydelse för företag.* ISSN 1102-8254.
[210] Huzzard,T. *The Convergence of the Quality of Working Life and Competitiveness. A Current Swedish Literature Review.* Stockholm: Arbetslivsinstitutet 2003:9.
[211] Sandberg,J. & Targama, A. (1998). *Ledning och förståelse, ett kompetensperspektiv på organisationer.* Lund: Studentlitteratur.
[212] Ahrenfelt, B. (1995). *Förändring som tillstånd.* Lund: Studentlitteratur.
[213] Herzberg, F. Mausner, B. & Snydeman, B. (1959). *The motivation to Work.* New York: Wiley & Sons.
[214] See e.g: Ekvall,G. (1991). *Förnyelse och Friktion, om organisation, kreativitet och innovation.* Stockholm: Natur och Kultur.
[215] Arvonen, J. (1989). *Att leda via idéer.* Lund: Studentlitteratur.
[216] Leithwood,K. Jantzi,D. & Steinbach, R. (2002). *Changing leadership for changing times.* Open University Press, Buckingham and Philadelphia.
[217] Trollestad, C. (1998). *Människosyn i Ledarskapsutveckling.* Nya Doxa.
[218] In Docherty, P. Forslin, J. & Shani, R. (eds.) (2002). *Creating sustainable Work Systems: Emerging perspectives and practice.* Routledge, London and New York. It is unfortunate, however, that they use the term reflexive rather than reflective, for it is indeed reflection which is involved.
[219] Abrahamsson, K. et al. (2003). *Friskfaktorer i arbetslivet.* Stockholm: Prevent.
See also: Kira, M. *Moving from consuming to regenerative work* in Docherty, P. Forslin, J. & Shani, R. (eds) (2002). *Creating Sustainable Work Systems.* London: Routledge.
[220] Egon Rommedal, psychotherapist, lecture at the managerial course at Alingsås Sweden 3 March 2004.
[221] WHO, *Ottawa Charter for Health Promotion* (1986) www.who.int.

[222] In Forsberg, E. & Starrin, B. (eds.) (1997). *Frigörande kraft.* Förlagshuset Gothia.

[223] Seligman, M.E. & Garber, J. (eds) (1980). *Human helplessness, theory and applications.* New York: Academic Press.

[224] Feeling of being alien i.e of not belonging . At work, brought about by a lack of information about and control of one's own work.

[225] P. 212 in Abrahamsson, K. et al. (2003). *Friskfaktorer i arbetslivet.* Stockholm : Prevent

[226] Svedberg, L. (2003). *Grupppsykologi, om grupper, organisationer och ledarskap.* Lund: Studentlitteratur..

[227] Starrin, B. & Jönsson, L.R. Ekonomisk påfrestning, skamgörande erfarenheter och ohälsa under arbetslöshet. *Arbetsmarknad & Arbetsliv,* 4:2 1998.

[228] Theorell,T. Är ökat inflytande på arbetsplatsen bra för folkhälsan ? Kunskaps sammanställning. Statens folkhälsoinstitut. 2003:46

[229] The full official title in translation is The Employment (Co-Determination in the Workplace) Act SFS 1976:580

[230] Karasek, R. & Theorell, T. (1990). *Healthy Work.* New York: Basic Books Inc.

[231] Sennet, R. (1998). *The Corrosion of Character, the personal consequences of work in the new capitalism.* New York: W.W. Norton & Company Inc.

[232] Rendahl, J-E. Hart, H. Lawler, E. Ledford, G & Norrgren, F. (1995*). Att förandra och leda morgondagens arbete.* Stockholm: Vis strategi AB.

[233] Their, S. (1996). *Det pedagogiska ledarskapet. Mariehamn*: Mermerus förlag.

[234] Maltén, A. (1981). *Vad är kunskap?* Malmö: Gleerups förlag.

[235] Included in Thelander, E. (2003). *Delaktighet och dialog- på väg mot hållbara arbetsplatser.* Stockholm: Arbetslivsinstitutet.

[236] Docherty, P. Forslin, J. & Shani, R. (eds.). *Creating Sustainable Work Systems: Emerging perpectives and practice.* Routledge, London and New York.

[237] Steinberg, J.M. (1994). *Den nya inlärningen.* Stockholm: Ekerlunds.

[238] In Docherty,P. Forslin, J. & Shani, R. (eds.) (2002*). Creating Sustainable Work Systems: Emerging perspectives and practice.* Rouledge, London and New York.

[239] In Lennéer-Axelsson, B. & Thylefors, I. (1996). *Om konflikter, hemma och på jobbet.* Natur och kultur.

[240] A psychological term which indicates the extent to which an individual has the capacity to show interest in "the other".

[241] Thelander, E. (2003). *Delaktighet och dialog- på väg mot hållbara arbetsplatser.* Stockholm: Arbetslivsinstitutet.

[242] In Thelander, E. (2003). *Delaktighet och dialog-på väg mot hållbara arbetsplatser.* Stockholm: Arbetslivsinstitutet.

243 In Docherty, P. Forslin, J. & Shani, R. (eds.) (2002). *Creating Sustainable Work Systems : Emerging perspectives and practice.* Routledge, London and New York.
244 Theorell, T. *Är ökat inflytande på arbetsplatsen bra för folkhälsan?* Folkhälsoinstitutet (2003).
245 Demmer, H. (1995). *Worksite Health promotion: How to go about it.* WHO Collaborating Centre BKK. ISBN 3-924379-40-8.
246 In von Otter (ed) *Ute och inne i svenskt arbetsliv- forskare analyserar och spekulerar om trender i framtidens arbetsliv.* Arbetslivsinstitutet, Arbetsliv i omvandling 2003:8.
247 Ibid (p. 380)
248 Huzzard, T. *The Convergence of the Quality of Working Life and Competitiveness* A Current Swedish Literature Review Arbetslivsinstitutet 2002:9 ISBN 91-7045-683-6.
249 To be found in Docherty, P. Forslin, J. & Shani, T. (eds.) (2002). *Creating Sustainable Work Systems: Emerging perspectives and practice.* Routledge, London and New York.
250 Ahrenfelt, B. (1995). *Förändring som tillstånd.* Lund: Studentlitteratur.
251 Watzlawick, P. Weakland, J & Fish, R. (1974). *Change; Principles of Problem Formation and Problem Resolution.* W.W. Norton&Company
252 Marmgren, L. Att leda mjuka processer, *MiLs Skriftserie* nr 17/2002. ISSN 1402-3466.
253 Ahrenfeldt, B. (1995). *Förändring som tillstånd.* Lund: Studentlitteratur.
254 Schein, E.H. (1988). *Process Consultation Volume I - Its role in Organization Development.* New York: Addison-Wesley Publishing Company.
255 In Docherty, P. Forslin, J. & Shani, R. (eds.). *Creating Sustainable Work Systems: Emerging perspectives and practice.* Routledge, London and New York.
256 WHO, *Ottawa Charter for Health* Promotion (1986) www.who.int
257 Seedhouse, D. (2004). *Health Promotion, Philosophy, Prejudice and Practice.* Wiley & Sons.
258 Sandberg, J. & Targama, A. (1998). *Ledning och förståelse, ett kompetensperspektiv på organisationer.* Lund: Studentlitteratur.
259 Marmgren,L. *Att leda mjuka processer,* MiLs Skriftserie nr 17/2002. ISSN 1402-3466.
260 Schein, E.H. (1988). *Process Consultation Volume I - Its role in Organization Development* . New York: Addison-Wesley Publishing Company.
261 In Docherty, P. Forslin, J. & Shani, R. (eds). (2002). *Creating Sustainable Work Systems: Emerging perspectives and practice.* Routledge, London and New York.

Printed in the United States
96089LV00006B/133-156/A